Nick Baker's
BRITISH
WILDLIFE

First published in 2003 by New Holland Publishers (UK) Ltd
London • Cape Town • Sydney • Auckland

2 4 6 8 10 9 7 5 3 1

Garfield House, 86-88 Edgware Road, London W2 2EA, United Kingdom
www.newhollandpublishers.com

80 McKenzie Street, Cape Town 8001, South Africa

Level 1/Unit 4, 14 Aquatic Drive, Frenchs Forest, NSW 2086, Australia

218 Lake Road, Northcote, Auckland, New Zealand

All Photography by David M. Cottridge, with the exception of those
listed on page 160

All artwork commissioned by Wildlife Art Ltd. www.wildlife-art.co.uk,
with the exception of those listed on page 160
Front cover illustration by Stuart Carter / Wildlife Art Ltd.
Main artists: Stuart Carter with additional artwork by Dave Daly, Sandra Doyle, Sheila Hadley,
Stephen Message, Sandra Pond and Mike Unwin

ISBN 1 84330 060 5

Publishing Manager: Jo Hemmings
Project Editor: Lorna Sharrock
Copy Editor: Sylvia Sullivan
Editorial Assistants: Daniela Filippin, Gareth Jones
Designer: Alan Marshall
Assistant Designer: Gulen Shevki
Cartography: Bill Smuts
Index: Janet Dudley
Production: Joan Woodroffe

Page 1: Bank Vole
Page 3: Nick Baker seawatching
Pages 4-5: from top left: Waxwing, Badgers, Daffodils, Honey Bee, Common Lizard, Blackberries,
Bluebells, Early Purple Orchid, Common Frog, Red Deer, Mountain Hare, Smew
Page 6: Nick Baker birdwatching

Reproduction by Modern Age Repro Co. Ltd, Hong Kong
Printed and bound in Malaysia by Times Offset (M) Sdn Bhd

The wildlife TRUSTS

Nick Baker's
BRITISH WILDLIFE

A MONTH BY MONTH GUIDE

NEW HOLLAND

Contents

 The Wildlife Trusts 7
Introduction 8

The Wildlife Trusts

The Wildlife Trusts partnership is the UK's leading voluntary organization working, since 1912, in all areas of nature conservation. We are fortunate to have the support of more than 380,000 members – people who all care about British wildlife – and, of course, vice president, television presenter and author, Nick Baker.

The Wildlife Trusts protects wildlife for the future by managing almost 2,500 nature reserves across the UK, ranging from wetlands and peat bogs, to heaths, coastal habitats, woodlands and wildflower meadows. These habitats are home to Britain's best-loved wildlife, from the Otter and Bittern, to the Basking Shark, Red Squirrel and Common Dormouse.

Some species are easier to spot than others and this book gives an excellent overview of what is happening in nature throughout the year. The month by month account covers mammals, birds, amphibians, reptiles, fish, invertebrates and plants.

With Nick Baker's help you will know when and where to start looking for wildlife, from Bluebell woodlands and boxing Brown Hares to a Badger family at dusk. Your local Wildlife Trust will have information on wildlife activities for adults and children, including a variety of Wildlife Watch events.

The Wildlife Trusts works with Government, planners, companies and the general public, to raise awareness of the importance of threatened habitats and the need to protect them. Importantly, we encourage people to 'do their bit' for wildlife. We believe that a deeper appreciation for nature conservation can start with a book such as *Nick Baker's British Wildlife*. We need more people to understand and value the wildlife that surrounds us.

Few people realize just how endangered much of our British wildlife is. In recent years once common species such as the Tree Sparrow and Song Thrush have declined. Hedgehogs, frogs and hares have also suffered – mainly as a result of the demands of modern living on habitats, for water, housing and transport.

The Wildlife Trusts believes that it is not too late and that we all have a part to play in reversing the losses of the past and ensuring that the UK really is a better place for wildlife and people.

Help us to protect wildlife for the future and become a member today! Please complete the form attached, or phone The Wildlife Trusts on 0870 0367711. Log on to www.wildlifetrusts.org for further information.

Thank you for buying *Nick Baker's British Wildlife* – and we hope you have fun discovering the wildlife on your doorstep!

The Wildlife Trusts is a registered charity (number 207238).

Introduction

Be aware that the topic I have tried to tackle in this book – wildlife in Britain and Ireland – is so vast and varied that, although it is a subject that I love and that has been a huge part of my life, my knowledge is riddled with many holes, especially in the world of those largely green denizens of the world – the plants! For any limitations in this area, or for any potential shortcomings of expectation, I apologize wholeheartedly.

This book certainly isn't intended to be a field guide to everything that walks, swims, flies or photosynthesizes here in the UK. Nor is it intended to be a simple coffee table book that sits on a shelf, unread, unloved and unwanted (why on Earth they are called coffee table books I don't know, they rarely ever spend much time there). My own coffee table is frequented by books I like, so having said that I guess I do want it to be a coffee table book after all!

I hope that this book will inspire anyone who reads it to look out of their windows, windscreens and binoculars in a different way and see things they have never seen before. After all, we live in a country that despite seeming sterile and grey at times, is actually far from it. Knowing where to look at what time is the secret to its beauty.

That being said, I do not wish to spell out every experience! (There is nothing worse or more patronizing than an idiot's guide to the world, complete with glaring signposts.) I want merely to start the ball rolling in the right direction for maximum enjoyment and future inspirations.

For example, just getting the timing right can make all the difference. A visit to a mud flat during an hour or so either side of high tide can turn the first timer into a wader watcher

for life, just by delivering all the best features up close and in their full gorgeous glory, doing all those interesting things that the field guides tell you about! Get it wrong, and all the enthusiastic amateur will see will be a few unidentifiable, animated specks in the distance. The disappointed may well turn to other pursuits and for the rest of their days believe that watching omnibus editions of *EastEnders* or even macramé are more stimulating ways to pass a wintry Sunday afternoon (take it from me, they aren't, I've tried both and they simply don't work!).

Part of the magic of wildlife for me is its unnerving ability to surprise you. Just when you think you have seen, heard and smelt everything about a place, something rocks up and knocks your socks off.

ABOVE: Rockpooling and beachcombing
Catch the tide when it is high and who knows what extraordinary creatures you may find in your net. Wildlife watching with friends is double the fun and they can help you carry any equipment you may need, as well as other necessities - lunch!

Tidal pools are just one such place that continue to surprise me. I've been turning rocks over ever since I've owned my own bucket and spade, in fact I feel like I've looked under every rock from Bangor to Bognor. Despite this, just the other day I was startled to find under a particularly unassuming rock in Torbay at least two of the funkiest little beasts that I had ever set my eyes on – a turquoise flat worm, straight out of Jimmy Hendrix land, and a white and orange sea slug, with all the grace and flounce of a 1930s lady's boa!

To sum up, the message behind this book is simply 'Get out there!' Wake up, smell the Primroses, fresh mown grass, guano, whatever, just do it!

ABOVE: Look, but don't pick
A vast swathe of Bluebells in a woodland in spring is enough to tempt anyone to pick them. But don't!

Field Etiquette and the Rules of Rambling

As much as I would love it, it isn't always possible to go skipping off into the woods and water meadows to investigate and explore any flower or patch of land, mountain or bog whenever the whim or wanderlust strikes us. We do have to face reality. We live on an overcrowded bundle of islands, much of which is owned by someone. There are barriers, boundaries and properties that must be respected and much of the real special places are set aside to preserve them as just this. The living things that share our country and the planet in general are already under massive pressure without hundreds of well meaning amateur naturalists, pointing, probing and trampling all over the place. But it is still possible to enjoy the wild – and not so wild – life of said islands by following the basic codes of country conduct.

Country code – Baker's interpretation
Here's my interpretation of the countryside code.

1 Shut gates after you. Simple one this, but many a farmer's day has been ruined by a few seconds of thoughtlessness. Also I know of many conservation project disasters caused by a herd of innocent herbivores, trampling and chewing their way through an endangered species or two or eating the habitat of another.

2 Keep dogs under control. It doesn't matter how well you think you know Rover. He is a predator at heart and can be driven to all sorts of mischief, from chasing and stressing of farm livestock, to the disturbing sensitive wildlife. It also doesn't help your own wildlife watching. There is nothing more frustrating for you and the wildlife than having an enthusiastic hound bounding around and sending all and sundry fleeing for what instinct says are their lives!

3 Keep to paths. They are there for a reason; not only do landowners get upset if you go bumbling across their property or crops, but they are also useful for relating your position to maps, for the quiet stalking of wildlife. In areas of impenetrable vegetation they enable you to move about freely and they act as a focus for wildlife too.

4 Use gates and stiles. Obvious this one – going through hedges hurts as can getting snarled up on barbed wire. There is also the potential for damaging hedges and fences.

5 Leave plants alone. By this I mean do not pick them. Sure, encouraging children to collect leaves and possibly a bunch of common flowers for Mothering Sunday, and samples to take home for tricky identification, are all well and good, but avoid uprooting armfuls of the things and do not damage endangered species.

6 Take litter home. That means the lot: apple cores, banana skins, anything that wasn't there when you arrived.

7 Be careful with fire. Matches, cigarettes, and barbecues can all start fires, especially during dry weather on grassland and heath. Fires can easily get out of control and damage huge areas of land and put life at risk.

8 Be careful on country roads. Cars hurt if you hit someone or something in one. If driving, I always find it's best to expect a hazard, like a horse or a tractor, around every corner. That way when there is a horse, you'll be ready for it!

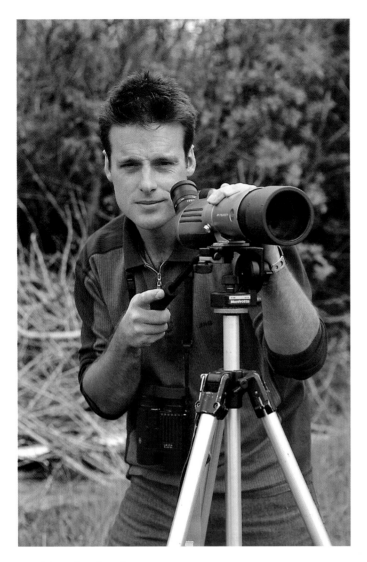

ABOVE: Blending in
Wearing muted colours whilst out in the countryside will mean you are more likely to see the shyer species.

Field tips, or how to minimize impact and get the best out of your wildlife watching

As much as I hate rules and regulations, there are a few things which can escape your notice and there are good habits to get into, especially if you are working with young and impressionable people. I'm not being patronizing here, just trying to be helpful. After all, I've been caught out myself on a few occasions.

All the toys, gadgets and knowhow in the world mean nothing if you sound like an army marching on crisp packets. The world we perceive is mainly through our sense of sight. Do not forget that animals' senses are often much more astute than ours. Their eyesight can be sharper but, even more importantly, their sense of smell and hearing can be hundreds of times better than ours. Wildlife is wild life and humans are predators by nature and everything knows that. Be quiet, move quietly and be rewarded.

1 Wear the right clothes. While I wouldn't want to extol the fashion virtues of dull greens and camouflage gear, there is a point to this stuff, especially if you are trying to get close to something particularly highly strung. However, any clothes that are not dayglo orange or red should be fine, the darker the better. The sensible use of field craft like taking advantage of natural cover, avoiding being silhouetted, and approaching downwind (for mammals especially) will do more for your wildlife watching than even the most expensive camouflage clothing!

Bear in mind that by wearing clothing that blends in you are really satisfying your own insecurities, those revolving around being seen. The reality is that vision is not the number one sense to most of Britain's most nervous animals, the mammals. (Birds, especially groups of them, are another ball game altogether and both sight and sound play an important part in their early warning systems.)

2 Observe, unobserved. I have walked up to quivering deer, downwind and quietly in full view, with the animals just staring at me. They can see me, but it's not until I confirm my presence to one of their more vital senses like hearing or smell, by treading on a twig or through the wind changing direction, that they freak out and flee.

Number one priority is be quiet. No rustly papers, no swishy clothes, no tearing Velcro. Keep potential snags like laces and straps well tucked away.

A good little exercise, especially with children, is to play a blindfold game. One person stands still wearing a blindfold, and the more obstacles like crunchy leaves, twigs, etc. the better. A second person then has to get from one point to another without the blindfolded person hearing them and pointing at the spot. Get pointed at three times and you are out. The 'stalker' soon learns to keep everything under control, even breathing can give the game away!

A few other tips are centred on the above. By understanding how animals perceive their world you have a good basis for a lot of fun and some close encounters.
- Try to keep numbers of people to a minimum especially in a strange situation. More people make more noise, whether they like it or not.
- Try to avoid strong smelling food, perfume and the like.
- Keep your eye on your quarry and move slowly to avoid treading on anything that will give the game away. If there is natural background noise, like wind in the trees, use this. Move when the wind blows and rustles the leaves, stop when it dies down and, if the animals you are trying to get close to look a little wary, just stop. Wait for them to relax and carry on.

Just by obeying these simple little rules, I have had many wonderful wild situations, all of which will be etched in my mind for ever.

3 Leave things as you find them. By all means have a rummage, but return rocks and logs to their original positions and release specimens where you found them. I have seen people on a field course replace a crab under a rock by putting the rock on top of the crab. Doing this, the chances are that the poor animal will be mushed! Instead, put the rock down and place the animal next to it so it can find its own way in.

4 Be gentle, tread lightly. This is as much for the animals' and plants' good as it is for your own.

5 Keep an eye on tides and timings. Always let someone know where you intend to be and roughly what time you will be back. It saves people panicking. Partners, parents and friends often do not quite understand. You might be having the time of your life with a baby badger sniffing your boots, while they are fussing around oblivious to your pleasure phoning hospitals and calling the emergency services!

The other thing I must stress here, because I have been victim to this one myself, is to note tide times. Rockpooling or birdwatching on the beach can be very addictive pastimes. Sometimes you forget how far you have travelled and before you know it several hours have elapsed, the moon has cast its influence on the waters and that headland beyond which is home suddenly becomes a cliff, with angry waters foaming around its base.

Getting trapped is not funny and can be life-threatening. At best you get soaked, having to swim home or scare yourself to death doing a solo rock climb (both are strongly advised against).

Tools of the Trade

Again, there are no surprises here! All real common sense stuff. Being a naturalist can be as expensive and gadget-ridden as you like or a very simple affair and, at the end of the day, it is good old-fashioned field craft which is the hardest thing to master. Any number of super Teflon-coated, rustle-free, self-ventilating jackets cannot make up for a bit of skill.

Having said that, there are a few items that seriously enhance the pleasure of being outdoors.

1 Binoculars. Even those of us with 20/20 vision will miss out on a lot without some kind of visual enhancement. A decent pair of binoculars are indispensable; buy the best you can afford – they will pay dividends. Models at the higher end of the market may seem a little excessive but they work so very well. The optical qualities of the lenses are second to none and probably one of the most important features for me at least is that they are waterproof, not just resistant, I mean rinse under the tap waterproof! This means you do not have to worry about them at all, and that makes the perfect tool.

2 Telescopes. These obviously do the same sort of thing but increase your range. They are, however, a luxury, unless you are doing something like seawatching, and will add extra burden in the form of a tripod and yet more weight. That being said, I love to be able to see the colours of a bird's bill or count its

ABOVE: Essential equipment
Whalewatching, birdwatching, even trying to find the ice cream van at the end of the beach, binoculars are invaluable when you're out and about. Invest in a good pair.

nasal hairs, and a telescope often accompanies me to hides and nature reserves.

3 Magnifying lens. Again another visual sense enhancing bit of kit. A good little one, which you can keep in a pocket, is relatively cheap. Essential for botanizing or looking at the creepier, crawlier side of life.

4 Notebook and pencil. Every naturalist will recommend them, but actually accost one and ask him or her to show their notebook to you and they will probably have left it at home. Notebooks are very useful, but they are just a habit most fail to get into. Keep a notebook and you will find it aids your memory, and will help you as a point of reference in the future or to identify something you saw but could not recognize in situ. They are also very handy for jotting down directions to nature trails, hides and pubs!

5 Torch. I'm not going to bother spelling out why these are useful. A little one, for the keyring or pocket, is very useful during the day, for shedding a little light into cracks and crevices especially when rock pooling.

6 Specimen pots and bags. You can never have too many of these, handy for keeping botanical specimens fresh and fragile things, living or dead – intact and unstressed.

RIGHT: Hand lens
To get up close and personal with all sorts of wildlife, get yourself a magnifying hand lens. Of all the gear you could own, this is one you cannot make and it is worth spending a little money on a good one.

January

Bloated by the excesses of the festive period and kept in by the cold, many miss what this month has to offer. During a cold snap food, becomes scarce or hard to access by animals that are still active and the drive to survive becomes stronger. Inhibitions are lost, the shy become shameless, the quiet become noisy and the secretive let their cover slip. Even the most sterile winter landscape will offer signs of life if you go out and stand still, stop rustling, stop puffing and beating your hands to stave off the cold, and listen. All will become clear and wildlife will seem to come to you.

ABOVE: Red Fox foot-prints in the snow
You can do more than just identify an animal by its tracks; you can tell what it was doing and when it was doing it, all without even see-ing it! This Red Fox was clearly running to or from something.

Mammals

Mammal detectives seeking small signs of life have a slightly easier job this month. With chance of snow, trails are going to be more obvious and like a big 'etch-a-sketch' the whole secret world is depicted as a story.

Uncovering voles
Wood Mouse tracks can look like miniature kangaroo or Rabbit tracks and will often confuse the amateur detective as it appears that the world is populated by three-legged mice. A bouncing mouse crossing an open area hastily will often leave its front prints superimposed. By looking at their spacing you can be forgiven for thinking they have wings as they seem to spend so little time on the ground between bounds.

But it isn't just footprints in the snow that you must look out for. Activities increase as their little bodies feel the squeeze of winter's grip; they get bolder, more adventurous moving farther afield. Feeding signs are often left in open places against a sparse backdrop of the naked countryside; 'nibblings' show up better.

Find a tussocky place, where grass has been permitted to grow lanky and rank, and pull back any grassy overhangs. Look carefully and you are almost certain to find the runs of **Common**

Field Vole tunnel – signs to look out for

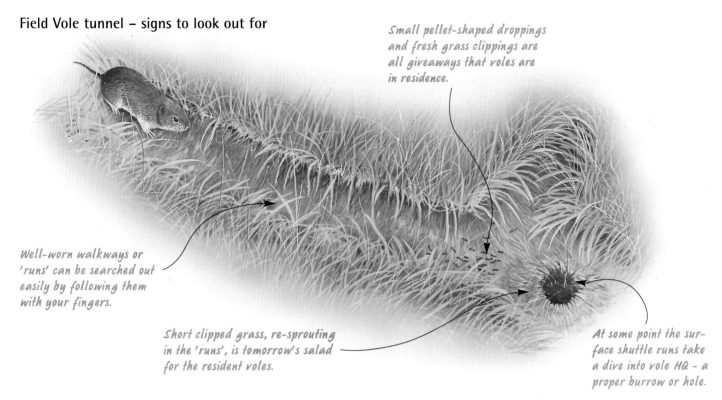

Small pellet-shaped droppings and fresh grass clippings are all giveaways that voles are in residence.

Well-worn walkways or 'runs' can be searched out easily by following them with your fingers.

Short clipped grass, re-sprouting in the 'runs', is tomorrow's salad for the resident voles.

At some point the surface shuttle runs take a dive into vole HQ - a proper burrow or hole.

and **Field Voles**. These will have been habitually defended and patrolled during the breeding season by one pair. Now they have become more tolerant of each other. Just finding enough to eat takes priority over sex in their labyrinth. Fresh piles of droppings in latrine areas and a carpet of grass clippings, or grass that has been shredded (look for short stems, that look like they have been cut by scissors and are not yellow at the edges) are signs that there are residents around.

Another very special mammal to be spotted is the **Water Vole**. For the same reasons their activities are easier to see as waterside vegetation dies back, but you have to assume the weather is good – flooding can prove disastrous for both vole and mammal detective – washing away field signs and even the animal itself. At its less extreme the weather plays an important role; Water Voles are fair weather mammals and hate the cold. So wait for a day with a bit of sun and walk the weedy waterways and you stand a better chance of, if not actually seeing the sun-loving aquanauts, at least hearing that reassuring plop as they 'dive' for safety.

ABOVE: Field Vole runs
Runs can be found anywhere that there is long grass that has collapsed on itself. Gently part the grass with your hands to reveal a vole subway and signs.

What to look out for this month

- Chasing squirrels
- Waders and wildfowl
- Raiding Sparrowhawks
- Early frog spawn
- Lichens
- Snowdrops
- Winter Heliotrope

LEFT: Water Vole
Look out for Water Voles swimming in calm water with lots of low grassy vegetation. They're much bigger than the more common Field and Bank Voles, which don't swim, and they have a more rounded appearance than Brown Rats.

Higher love

As we walk through parks and woodlands there is a common sound that never fails to attract our attention. It is such a head turner because it is loud and sounds unlike any other animal voice. This wheezy barking, often described as 'chuk-chuk-chuk-aaahh' is the alarm call of one of our most pugnacious mammals, the **Grey Squirrel**. By January the hormones are pumping around their little furry bodies and they are getting quite frisky. They have an uneasy relationship with humans which probably stems from our guilt in having introduced them to this country from North America in the first place. As they spread they have wreaked havoc on forests and on populations of the native Red Squirrel. However, easy though it is to get on our biodiversity high horses, let's face facts. It's too late, they're here and they are one of the few animals that just about everyone can see easily, every day. So enjoy them!

Now is a good time to start developing a crick in your neck as squirrel activity is never more noticeable than it is now. Squirrels can be seen courting in spring and autumn too but the peak of the action is this month which is handy as there are no leaves on the trees and they can be easily followed as they go about their three-dimensional aerial kiss-chase.

The courtship chases themselves are fast and furious. They are accompanied by much scrabbling, tail flicking and beating of the feet on branches and a very distinctive staccato kissing bark, which is issued by the males. It will take a little bit of time before you develop a feeling for what is going on as you follow these furry tornadoes of passion as they stream through the branches of a wood. The female develops an irresistible perfume which attracts the males in from downwind. She then leads an entourage of up to eight males on a merry dance through the trees. The males tear after her and attempt to woo her and defend her honour against the others. Eventually the most athletic male mates with the female.

You can sometimes lure single squirrels quite close or even initiate a bout of alarm calls by keeping absolutely still and making a kissing sound against the top of your fist.

LEFT: Courting squirrels
Follow the leader. She's the one in front - the males chase and scurry after her in a sorry display of desperation. But that's squirrel sex for you and a sign that the breeding season is upon us.

Birds

'Ooooohing' as a flock of **Knots** rush through for the first of many fly pasts before they eventually come to roost, it is easy to forget what a dilemma many of these small wading birds face.

Super-organisms

When you get a mass of birds together they seem to develop a group consciousness, they become like cells in a super-organism that is the flock, seething, wheeling and changing direction like some huge mobile abstract sculpture. Together they produce a spectacle that is far more than the sum of its parts.

This idea of a super-organism isn't too far from the truth. Throw a bundle of birds together and lots of interesting things happen. For a start there are more eyes – a flock of a thousand **Dunlin** has two thousand eyeballs peeled for the same danger that would face a single bird feeding on its own. This is bad news for a **Sparrowhawk**, **Merlin** or **Peregrine** as their sneaky manoeuvres to launch a surprise attack are likely to be rumbled in the early stages, sending the birds off into a confusing flurry of wing-bars, tail stripes and aerial stunts.

Many eyes mean more food

The hidden benefit for the flocking birds here is also one of time spent feeding. This can be illustrated if you watch another winter aggregation – geese grazing. Heads pop up from the flock at regular intervals – these birds are the sentinels, frequently checking for danger. It has been shown that the number of 'peepers' does not increase in proportion to the size of the flock –

it doesn't need to – the bigger the flock the less 'peeping' is required for each individual and the more valuable time to be spent feeding!

Birds are also thought to exchange information on where the good feeding is to be found. At a **Starling** or wader roost displays from individuals, their time of arrival or some other communication device are thought to relay information to others that had less successful foraging. The next day these birds may follow those which fed well. In this way the whole flock benefits by being able to exploit patchy resources, such as a ploughed field, or a bird feeder in a garden. Also watch waders feeding on the mud, Starlings on a lawn or the sombre forms of **Rooks** waddling across a field and this idea of the group benefiting can be seen in another way. When one bird hits a hot spot – perhaps a natural hollow out of the wind where worms may be closer to the surface, it starts pecking and in doing so alerts its feeding companions. Once the patch has been cleared they spread out again. Lots of eyes and ears means that a systematic and efficient sweep of a habitat can be made for food.

Snuggle up!

The other important benefit of flocking is simply the conservation of body heat – many birds spend the entire winter period carefully balancing the amount of energy consumed and the energetic cost of heating their bodies. Many simply seek cosier and warmer locations over water in reedbeds, a favourite location for Starlings.

ABOVE: Waxwings
Waxwings are regular winter visitors to the east of Britain when the weather on the continent becomes simply too cold. These gorgeous, plump little glam rockers of the bird world can be quite tame and flock to towns and gardens to feed on berries.

BELOW: Lapwing flock
Like broken green bottle glass – there are few finer sights than a flock of Lapwings feeding on a waterlogged field in the watery winter sun.

Top Muddy Places

Clyde Estuary, Strathclyde (NS390730 and NS315785) – Probably one of the most accessible of the Scottish estuaries, not far from the M8. The south shore is rather more built up, but good birding is still to be had. The north shore is a mix of mud, saltmarsh and rocky shore and this is reflected in a good mix of wildfowl and waders, including Turnstones, Dunlins and flocks of swans and Pink-footed Geese.

Greyabbey, Co Down (582672) – The abundance of wildlife in and around Strangford Lough is a delight. Just south of Greyabbey on the Ards Peninsula it has permanent residents such as Redshank, Oystercatcher and other waders. Great flocks of Brent Geese, thought to represent 75% of the world's population, visit here too.

Morecambe Bay, Lancashire (SD468666) – Every naturalist should visit this at least once. It is big on mud, big on sand, big on molluscs and worms, and big on birds. A quarter of the country's wintering Bar-tailed Godwits, Knot, Turnstones, Dunlins and Oystercatchers can be found here as well as a rich and varied waterfowl population.

Exe Estuary, Devon – This is one of the most productive estuaries in the west country with a good range of habitats and locations to watch. It is also rather user friendly being spread out over a relatively small area. Good spots to visit for huge high tide roosts of waders are Dawlish Warren (SX983794) and Bowling Green Marsh, Topsham (SX958870). Large flocks of 300 or more Avocet are a particular highlight near the mouth of the River Clyst at Topsham.

Snettisham, Norfolk (TF648320) – This RSPB reserve is a must. Slap bang in the middle of The Wash, it is a great place to visit whatever the weather. High tide can provide huge and stunning roosts of waders. The Knots can be so numerous that at a distance they appear to be one huge shingle bank. It is only when the tide retreats allowing access to the vast mud flats, that they blow their cover and waft out over the mud like smoke. There are plenty of other species, including a lot of excellent ducks that congregate on the shallow pools at the top of the shore allowing close views of species that normally raft well out at sea, not to mention the Brent and Pink-footed Geese coming in at dusk – awesome!

Old Hall Marshes, Essex (TL958118) – Like any of the cracking east coast estuaries, it gets very cold, taking the full brunt of any easterly wind. But this RSPB reserve is famous for a good mix of wadersand wildfowl with just about everything worth focusing your bins on this winter present, some in good numbers too. A bit of research beforehand is essential to avoid low tide and early morning can be difficult looking into the low sun. So, read-up, wrap up and enjoy!

Pagham Harbour, West Sussex (SZ856966) – This beauty is often over-looked and over-shadowed by the reputation of nearby Chichester Harbour (SU763003) and Langstone Harbour, which together form the largest single piece of mud in the south. But Pagham is just as good and is often more sheltered from the winter weather, and more pleasant for birdwatchers.

ABOVE: Avocet *More white than black, in winter they mass together. The name is from the French 'advocat', meaning barrister, refering to the black and white dress.*

ABOVE: Oystercatcher *More black than white, they never fail to cheer up the dullest of winter days along our shores. Listen out for their noisy piping calls.*

LEFT: Seawatching
I know there's a Little Auk out there somewhere! Stare at even the emptiest-looking ocean for long enough and you will see something, I promise! It may not be what you expected, but lots of surprises can turn up. A telescope is very handy.

BELOW: Seabirds at sea
When sea gazing, expect quite a variety of birds and identify birds by shape. A field guide is essential for beginners. Here we have Long-tailed Ducks, a Great Crested Grebe, Guillemots, a Red-throated Diver and a flock of black dots on the horizon!

Others roost on street lamps, bus shelters, railway station roofs, or trees outside restaurant extractor fans if you are a **Pied Wagtail**.

Small bodies have a larger surface area in proportion to their volume and so they lose heat quicker than a big one would. Even if it is just for a night, birds can find it advantageous to snuggle up. In cold weather **Wrens** and **tits** will roost in nest boxes. Roosting waders, such as **Redshank**, may not huddle in quite the same way but they will benefit from the presence of others. They all face into the wind to minimize ruffling their feathers but it is the adults that will take up the most sheltered positions, whilst the juveniles bear the brunt of the wind.

'Gaze the grey'

'Seawatching' is regarded by some to be a little bit too hard-core for good health! I have heard of extreme enthusiasts lashing themselves to promontories to hold themselves steady in the wind and spray. However, it doesn't have to be the macho end of ornithology that it is often cracked up to be – it's a top way to see a variety of birds that are rarely seen any other way and if you get your weather forecast right it needn't be too uncomfortable either.

Many northern breeding seabirds get frozen out of their summer breeding grounds and head

RIGHT: Shag
FAR RIGHT: Cormorant
When seawatching, the birds to separate first from the pack are the Cormorant and its close relative the Shag.

BELOW: Long-eared Owl

Big ears – one of our least-known owl species, which is best seen in winter. The elusive Long-eared Owl forms secretive winter roosts around the country. If you see an owl hunting by daylight over salt-marsh or grassland, then it's Short-eared, not Long-eared.

south for the winter as even the nastiest stuff that the British climate throws at them is balmy compared with the conditions of the Arctic and sub-Arctic. Ducks are often the most obvious birds as they form flocks. Look out for disorganized rafts of **Eider** with their long bodies, short neck and triangular 'door stop' profile as these are probably one of the commonest marine ducks. Others frequently seen are the spiky-headed **Red-breasted Merganser** and the **Common Scoter**, a cracking black Russian beauty that forms fairly large flocks. They are always worth sifting through as they are sometimes joined by a rarity such as the similar **Velvet Scoter** and my favourite, the elegant glamour boy of the tumultuous seas, the **Long-tailed Duck**. The male really looks out of place miniaturized by a big swell.

It is possible to see all three species of divers together off British coasts during the winter. They are all similar in appearance, being low-slung birds with dagger bills and a dive that is smooth and without the little jump start of **Cormorants** and **Shags**. The **Red-throated Diver** is small, pale with an upward tilted bill; the **Black-throated** is darker in body and neck with

white flank patch; and the **Great Northern** is a big, dark heavy-billed bird with a lot of dark on its cheeks. But take it from me – just as you are about to reach a decision on a bird's identity, they will always live up to their name and vanish below the surface, requiring you to start the process of elimination all over again when they return some 30 seconds later.

If you find the divers hard work, do not even attempt to identify the four species of **grebes** in winter plumage! They can all look irritatingly similar to each other and are guaranteed to have you ripping the pages from your field guide in frustration.

Expect the unexpected

It can happen to anyone anywhere this month! I remember a January walk when, immersed in my own thoughts, I walked through a gap in a Hawthorn bush to suddenly find two large amber eyes staring back at me. I jumped back and there, not a metre from the first, another pair and another. In the split second it took for me to regain my composure and pop the lens caps from my binoculars the three **Long-eared Owls** had scuffled through the underscrub and were winging their retreat along the hedgerow, unseen again by me.

Two lessons learnt: never have lens caps on binoculars (throw them away immediately!) and prepare yourself for anything whilst out and about in the winter.

A number of birds of prey can appear in places well away from their usual breeding haunts this month. As far as Long-eared Owls were concerned I was never to repeat that particular

experience, although I now know of several winter roost areas, which seem to have the favourable qualities of plentiful hunting habitat – wood, hedgerow and ditch edges – combined with shelter, normally dense scrub. The weather needs to have been bitterly cold and even roosts of ten birds can be pointed out to you and you can still leave the site without having set eyes knowingly on a single owl.

Cold conditions and oodles of luck are the magic ingredients increasing the chances of seeing drifting birds of prey during the winter. Rough grasslands, moor and saltmarsh always have the potential of turning up **Short-eared Owls**. Whilst well away from their northern breeding grounds, you have the chance of witnessing the slow, soft, rowing flight of **Hen Harriers**. You may see them as far south as the Channel coast in habitats from marsh, reedbed, downland and the surrounding arable land, where they often stray to hunt wintering flocks of feeding passerines on stubble and kale fields. **Barn Owls** can also be spotted, hunting low over fields for small mammals such as voles and rats.

Swan throng

Winter water rises, falls, ebbs and flows and fluctuates a lot. Flooding is frequent in January and rivers burst their banks and fields become lagoons – good stuff if you are a highly-mobile semi-aquatic vegetarian, which is precisely what swans are.

At this time of the year it is commonplace to peer out over a farming landscape and see what appears to be white flotsam washed to the edges of the waters.

Look closer and you will find they are one of the three British species of swan. The most common is the easiest to recognize as it is a familiar resident of the duck pond. The **Mute Swan** is the one with the orange on its beak. If its usual pond freezes up in a cold spell, then it will turn up in other locations where the water is flowing or the ice is thinner and will graze on both submerged plants and grasses.

The other two species are migrants, coming here for the winter only. They both tend to frequent arable areas near water, especially old stubble fields where they will Hoover up spilt grain or the occasional lost potato. They are also characteristically much noisier than the Mute Swan, which despite not being at all mute is definitely the least vocal of the trio.

The **Whooper Swan** arrives in variable numbers depending on how bad the weather has been in Iceland. It is about the same size as the

ABOVE: Mute Swans
Mute Swans are resident and are our commonest species of swan. As they are floating vegans, you can expect them to bob up just about anywhere there is water and flooding. The name mute refers to their relatively limited range of vocalizations compared with other swan species.

RIGHT: Whooper Swans
Whooper Swans are about the same size as Mute Swans. With the yellow on the bill, their long necks and wings, they are also very elegant beasts.

Mute Swan and has a yellow and black bill. This is where the confusion starts because the other migrant swan, the **Bewick's**, also has a black and yellow bill, with a little less yellow than its doppelgänger, but it has the distinction of being a much smaller goose-like bird with a shorter, thicker neck.

Blood on the snow
Many raptors, those hook-billed assassins, do their predatory work in woodlands. Their murderous activities are occasionally noticed by

owners of bird tables, when like a tornado they strike, sending the lucky scattering like feathered debris. The **Sparrowhawk**, somewhat surprisingly given its status, is rarely noticed by us.

It is probably our second most common bird of prey (around 32,000) after the **Kestrel**, but rather than hanging around where we can see it such as above road verges, the Sparrowhawk flies among us unseen, a bird of dense cover, bush and woodland. It used to be a good bet that any small raptor that you saw over a town would be a Kestrel, but Sparrowhawks are

RIGHT: Bewick's Swan
Bewick's Swans are stumpy and more goose-like than our other two swan species. They also sit higher in the water than their counterparts.

increasing in number and moving into more sub-urban and urban areas.

During the winter the lack of leaves makes Sparrowhawks easier to observe. I find, by walking in deciduous woodland and larch plantations where there is a high tree density with a spacing of 2–3 m and, by spending a little more time than feels natural looking skyward, you increase your chance of sighting this wonderful woodland raptor.

Its long tail feathers and short rounded wing profile are designed for high manoeuvrability and short, sharp hunting bursts, able to penetrate dense foliage.

The sparseness of winter vegetation also makes it easier to pick out habitual 'butchering blocks'. Prominent gnarly trees or stumps and fence posts serve the purpose; fresh blood, feet and feathers are their macabre decoration. If you are still having trouble making a date with the bird of the golden scowl then track down your nearest roost of small birds such as Pied Wagtails or Starlings and stake it out as they come in at dusk. It will then only be a matter of time before you meet a Sparrowhawk.

The female Sparrowhawk is a larger, browner bird than the male, which is slate-grey above with rufous-barred underparts.

The rounded, short wings and the long tail for manoeuvrability make the Sparrowhawk ideal-ly suited for short, sprinting flights through dense undergrowth.

LEFT: Sparrowhawks
The Sparrowhawk is making a controversial comeback and is now one of the most com-monly seen birds of prey. Be proud if one comes to your garden feeder to snatch a Blue Tit or two!

LEFT: Common Crossbill
In conifer plantations and pine forest keep an ear open for a high-pitched 'chip chipping' call from the tree tops. This, combined with a crunching sound and a rain of cone fragments, reveals a feeding party of Common Crossbills, harvesting the resource that allows them to get a head start and be the first birds of the year to breed.

Amphibians, Reptiles and Fish

RIGHT: Common Frogs go a-courting
If frogs are not yet spawning, the males will certainly be accumulating as a quick squint with a torch at night will reveal. Common Frogs 'purr' and do not go 'rrribitt!' as we have been misled to believe since early childhood!

I always used to spend time looking under stones and logs to see what animals were sheltering there. This was fine until my mother caught me destroying her rock garden! It was then I found a bit of corrugated sheeting, the type used as a roof for sheds and outhouses. It had been lying in a field for ages and on turning it over I found countless beetles, a **Slow-worm**, three **newts**, a **toad** and a vole complete with its nest. I had the idea to put down more sheets in other places to encourage wildlife – it worked. Scraps of carpet work just as well.

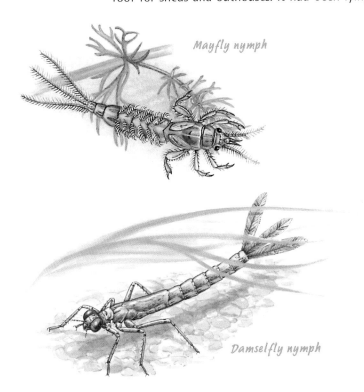

Mayfly nymph

Damselfly nymph

Invertebrates

If you need some signs of life in what, after all, can be a bit of a barren month, get in the water. For us with warm blood anything below 37° C is going to seem a little nippy, but for small invertebrates, being in the drink makes a lot of sense.

Water babies
Water, a poor conductor of heat compared with the air, acts as a thermal buffer so life beneath it can carry on pretty much oblivious to the hardships above the surface.

TOP LEFT: Mayfly nymph
LEFT: Damselfly nymph
Go and dip your net into your local river or stream and who knows what intriguing baby invertebrates you could turn up.

Nowhere is this life so diverse as in our rivers and streams. Dip a net into fast riffles and rapids and you may end up with nothing but a handful of gravel. Do it right by standing in the shallows upstream from your net, kick the bottom substrate a few times and you should get a result. Current in a stream slows towards the bottom and a thin boundary layer between 0.5-0.1 mm above the bottom is almost stationary. Thus most of the life to be found there is hanging on quite literally to the bottom.

Among the assortment of oddities now writhing in your net will be gadgets and gizmos a certain Mr. Bond would be proud of. Some, like **damselfly nymphs**, rely on brute force and tiny grappling hooks on each leg. Others will never get big enough to leave the safety of the crevices, such as the little two-tailed larvae of **stoneflies** and **mayflies** (three tails and a rounded body) belonging to the **Beatis** family. Other mayflies can be found in the fastest riffle, because they have a body shaped and designed to hug the contours of a pebble, not even raising an antenna to risk whipping them from their holdings.

Check out the extreme adaptations of *Ecdyonurus* and *Rhithrogena*. These two families look like they've been steam-rollered. Even their leg segments are flattened, giving them a hydrodynamic profile that works in the same way as the shape of a formula one car, creating negative lift and forcing the insect's body

down. As well as being kitted out with claw-tipped tarsi the latter group has modified gills which form a sucker.

To contrast, dip a net through the silty regions of a stream and you will find members of the same groups of insects but more tubular in cross-section, designed to burrow in the sediment. Some like the family of large mayflies that include *Ephemera danica* have gills on their back so they can breathe and be buried at the same time.

Beach balls

Beachcombing after a gale is profitable this month. Look for the jewel-like **Blue-rayed Limpet** on the holdfasts of Kelps. They sit in little 1 cm long depressions they have eaten away. They may have been responsible for their host's beaching.

Rock pools come alive with activity and there are plenty of eggs to be found here too. A wealth of **Dog Whelk egg cases** can be found clinging to the sides and crevices of overhanging rocks. Turn over rocks near the low tide mark and you may just be lucky enough to find a male fish guarding a cluster of eggs. The Common Shanny, the large and bloated Lumpsucker, the Common Goby, as well as the weirdest British fish, the Cornish Sucker, will all be nursing their clutches. Mysterious-looking bright green spheres are often stumbled upon now, attached to rocks and weed – these are the **eggs of Green Leaf Worms**.

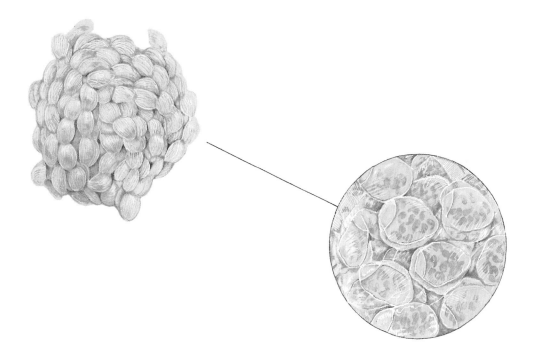

LEFT: Common Whelk egg cases
After a stormy sea, look out for 'Sea Washballs' - the egg masses of a spectacular sea snail, the Common Whelk. Each is a mass of weird flattened egg cases. Each individual egg case can contain thirty or more eggs, but only a very small proportion of these survive as they often become dinner to their siblings!

Plants

Orange-yellow splodges like lemon curd that's been hurled by the spoonful at walls and grave stones? Grey crusty stuff on rocks stuck with red-tipped hatpins? Old Man's Beard clippings neatly pegged along the twigs of an upland Oak woodland? This is the world of **lichens**.

With the leaves having departed the trees and the air damp, this is as good a time as any to look them up. The diversity gets better the further away from city centres you get, but even in our towns and parks you can find a good range of pollution-tolerant species – in fact lichens are so sensitive to air pollution they are used as a kind of environmental litmus paper to gauge air quality.

Did you know . . . ?

Lichens are decidedly odd. Actually they are fungi mainly of the order *Lecanorales*, but they differ from the rest of the mouldy world in that they have entered what is technically known as an obligate symbiosis with an algae of some form. The tiny plants cannot do without the protection of the fungus and the fungus cannot do without the carbohydrates or nitrates produced by the algae.

ABOVE: Heavenly Heliotrope
If you can smell vanilla, you could have discovered a clump of flowering Winter Heliotrope. This garden escape can be found all over Britain and its pink, spiky flowers are often confused with the later-flowering Butterbur.

RIGHT: Lichens
Visit your local grave-yard and you'll find plenty to keep you going all winter with around 1500 different species of lichen found in Britain.

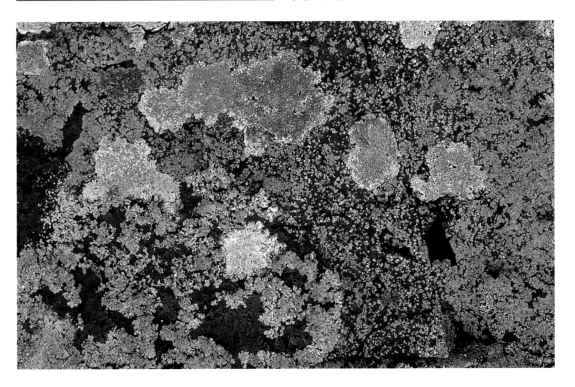

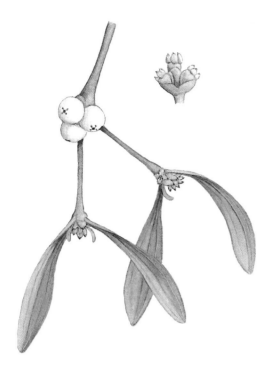

is an evergreen woody parasite that grows on other trees. Find a suitable tree in your garden or park – ideally it should be an Apple or Poplar, although it will grown on Hawthorn or Willow too, so you may have to brush up on your tree identification skills as well.

To plant a Mistletoe plant you need to make sure the seed stays on the branch. In the wild it is usually wiped on the twig by the bird that ate the rest of the berry and because it is so sticky it is glued in position. Many seeds are probably subsequently washed off by rain or drop off by accident. So to make sure your Mistletoe has every chance of germinating you need to use a trick (see below).

LEFT: Mistletoe
Mistletoe is a semi-parasitic plant on a variety of trees. It sinks its own roots into its host to acquire some nutrients and produces the rest by photo-synthesis, hence its green colour.

Christmas decorations

It's the time of year when the Christmas decorations are traditionally packed up and the holly, ivy and the tree all end up on the fire or compost heap. If the clump of **Mistletoe** that hung in the hall still has berries on it, now is the time to rescue them. If they are still firm and tacky they are perfect for a little project.

This doesn't always work so the more berries you can lay your hands on the better. Mistletoe

Practical bit

Grow your own Christmas decorations
Make a little split in the bark of the tree (preferably near a fork), so you can see the sappy wood beneath, then squidge the Mistletoe seeds into the gap.

Take an old strip of bandage and using one turn cover the seed and tie it securely. Keep checking it over the next season for any sign of a shoot and with a bit of luck you could be seeing your own Mistletoe plant – the ultimate in Christmas decoration recycling.

LEFT: Snowdrops
The terrestrial galaxies of Snowdrops are a mystery. Few actually produce seeds, but rely on fission of their bulbs and, despite hardened growing tips, are not native to the cold European countries.

February

February can be like meteorological roulette. One minute you can be tuned in to a 'Storm Cock', that is a Mistle Thrush, gushing forth his ballad in the sun, the next you can be running for the cover of a copse while being pelted by ice, snow and a wicked wind. Despite the seasonal peaks and troughs, there is a lot going on. It is unashamedly birds this month! They are the most obvious signs of life – not muted, frozen or suppressed by the cold and not as subtle as our limited mammal fauna. The thrum of insect life is hard to come by, unless we get an unseasonally warm sunny day.

Mammals

ABOVE: Storm Cock
I like Mistle Thrushes. When it's cold and blustery – the tops of the trees are swishing in turmoil and you pull up the collar on your coat and think wintry thoughts – this bird defies all, with a liquid ballad. An optimist, the Mistle Thrush is an admirable bird that sings for spring, whatever the weather.

Badger watching on a cold February night may seem like a precursor to going to see a psychiatrist but there are many advantages to watching our fossorial friends this month. There are no biting insects, the badgers emerge early (around 7–8 pm in undisturbed localities) and the drama that can be seen at the sett rivals that of a soap opera.

It is the main mating season, which means there is a lot of territorial activity. This is accompanied by a lot of bickering, purring and growling. You may even see the adult boar reinforce his post by scent-marking the sows, with a kind of 'eau de Badger', which is distinct to each social group.

Below the surface the sows are giving birth (watch for bedding being collected, this often reveals the position of the maternity den) and even though you may expect this to be a warm and maternal moment in the calendar, many subordinate females will lose their litters, probably to bullying boars or the infanticidal dominant sow.

Breeding
Visit your local Badger sett and you should start to notice signs of activity. Freshly ploughed out spoil heaps, old bedding and traces of new fresh bedding often dragged considerable distance to line the sleeping chambers below. An increase in activity coincides with the start of the Badger's year. Most cubs are born below ground this month.

Badger sett – signs to look out for

Badgers being large, flat-footed creatures of habit, weighing between 9-18 kg, any amount of coming and going will wear distinctive highways in the vegetation.

Setts are used over generations, so large spoil heaps of soil accumulate outside each entrance.

Badgers dig their setts into hillsides, where the soil is soft and free-draining and where there is plenty of undergrowth cover.

Trees near Badger setts, often Elder, are used for a quick manicure. By running their claws down the bark Badgers remove clogged-up soil from their toes and pads. These scratching trees are easy to spot.

Look for the long back hair caught up on wires where Badgers have pushed under fences. Hairs have a white tip, a black waist and a yellowish-white base. They are stiff and wiry: you can see why they were used to make shaving brushes.

Badgers are very clean - their world is ruled by their nose. Their latrines are more than a toilet; they are smelly signposts, used over and over again around the sett and on territory boundaries. They tell other Badgers who has been where, quite literally!

Badger footprints look very round with four of the five toes registering, distinctive long claws on the front feet and, when walking, the rear foot superimposes on the front.

ABOVE: Badger sett
The HQ for a clan of Badgers is unmistakable once you know what to look for.

RIGHT: Badgers
Look for signs that a Badger sett is in use - piles of bedding (dried leaves and grasses), scrapes in the soil for latrines, and footprints.

What to look out for this month

- Badgers breeding
- Heronries
- Ducks in their finery
- Geese
- Grebes in courtship
- Early butterflies
- Catkins

Birds

Collision of bill and resonant wood in a rapid staccato is one way two of our three species of woodpeckers advertise their claim on their wooded territories. Listen out for the loud percussion of **Great Spotted Woodpecker** and in comparison the almost whispering version provided by the diminutive and often overlooked **Lesser Spotted Woodpecker**.

RIGHT: Great Spotted Woodpecker
Great and spotty - the male Great Spotted Woodpecker is one of the first to herald the coming spring by hammering out his territorial rights on dead and resonant trees.

Rookery cacophony

Rookeries are worth watching. There will be lots of flapping to accompany the cacophony. Watch out for birds flying in with distinct bulging throats full of food, a sure sign that chicks have hatched in the high nests.

Dancing grebes

To actually witness the courtship of the **Great Crested Grebe**, the most over-produced, intricately embellished choreography you are ever likely to see, leaves you lost for words. Here's the best bit though, Great Crested Grebes are quite common birds; look on most large reservoirs and lakes and you should at least see a pair, and often more.

This handsome bird is easily spotted by its characteristic low-slung appearance, outrageous headgear and rather aloof manner, with its head held high as if it can't be bothered with anything but its vanity.

But bother they do, just watch them. Great Crested Grebe courtship is a stuttering performance that, if you read the books, occurs in five acts. The reality is that these acts do not necessarily follow in sequence. They are not discrete entities and to see one act certainly does not mean the others will definitely follow. The trick is to keep an eye out for any interaction between two birds, even if all they are doing is swimming towards each other. It is next to impossible to sex them, and both male and female play equal roles in courtship so you could well

BELOW: Lesser Spotted Woodpecker
Look out for these woodpeckers in woodland and parks. They are less likely to visit your garden than Great Spotted.

be about to witness a scene. The simplest manifestation of love is a variety of calls that advertise the fact that they are available. The next stage can be anything from facing each other to synchronized swimming toward or away from each other, or diving simultaneously.

Things begin to get spicy when there is a form of stylized preening, where both birds go through the actions of preening back feathers, often accompanied by head shaking, at its best an act that looks like the bird is dancing with its own reflection. There is also the 'cat display' which involves a lot of mewing while facing each other, accompanied by fanning out the tippets, those extraordinary flanges of feathers either side of the head.

But whatever order these early acts come in, the undisputed sign that passion on the pond has reached its pinnacle is the 'weed display' also known as the 'penguin dance'. It's an occurrence which you can never predict. It starts simply enough with the two birds swimming away from each other, diving and then rising out of the water together, legs paddling furiously as, belly to belly, they offer a symbolic beak full of nesting material in the form of pond weed. This is not exactly proffered so much as swung from side to side.

This is elaborate behaviour that potentially anyone can witness. In reality it takes seconds but lasts for ever in your memory. Just be prepared to watch a lot of grebes for a long time – there are still keen birdwatchers I know who have not yet seen the grande finale for themselves.

Courting Great Crested Grebes

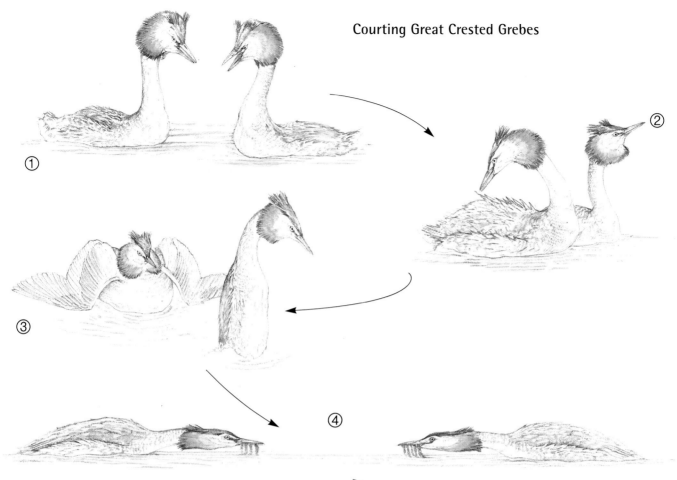

① ② ③ ④

ABOVE AND RIGHT: Dancing grebes
Look out on ponds and reservoirs for 'ball-room dancing' Great Crested Grebes as they follow their elaborate multi-staged courtship.

1 Pair face each other and shake their heads, often accompanied by neck arching and diving under each other.

2 Heads are turned away from each other in a bashful manner. This is often followed by ritualistic preening.

3 Female engages in a 'butterfly-like' posture, while the male stands erect.

4 After both have dived for a beakful of weed, a token nesting material, they face each other with heads low.

5 The climax to the prenuptials is the 'penguin dance'. Both birds tread water, rise to meet each other in a flurry of spume, beaks full of weed while shaking their heads in an exaggerated manner.

⑤

Ducking out

Take a winter saltmarsh. It wouldn't be that classic winter scene if it wasn't for ducks. A small group of dabbling **Teal** on the edge of the water, the rising whistles of **Wigeon** as more riffle in to join their number. The thing about ducks is you can see very different ducks in different places,

Check out a collection

A well stocked collection of tame birds can soon have you whistling through the identification problems associated with a distant duck in the wild, particularly the more subtle plumage of the females. For example, a quick lunchtime sandwich on a park bench in Regent's Park and by the end of the second round you could have notched up or even shared your sandwich with **Pochard**, **Teal**, **Wigeon**, **Gadwall**, **Tufted Duck**, **Eider**, **Goosander** and even a **White Pelican**!

The real thing

Of course, this is no substitute for watching the real thing in the wild, but it is a really accessible way of appreciating the broad-billed fraternity close up and provides unique views of those totally outrageous metallic shimmering sheens that bedeck the heads and wings of many males.

For those who need convincing that there is colour in February, look no further than your local gravel pit or village pond. Metallic greens and blues, marigold-yellow, glossy black and carrot-orange are just the colours found on your average male **Mallard** alone! There are at least 17 different kinds of duck that can be easily seen at the moment. In good oblique light it is hard to match the deep iridescent green of a drake Mallard's head or a Teal's face mask. Even the Tufted Duck, which you can be forgiven for describing as black and white, leaps into the royal realms of deep polychromatic purple at close range. The speculum, the band of bright colour in a duck's wing, also shines as if illuminated from within: green for Wigeon and Teal, deep blue for Mallard and bottle-green next to sky-blue for **Shoveler**.

Courting colours

Ducks are in tip-top condition now for a reason. All this showy, glowing plumage is a prerequisite for the breeding season that is just around the corner. Another good excuse to get into ducks this month is that there are no confusing juveniles or adults in moult.

With a little weak sunshine to warm their ardour, the beginnings of breeding behaviour can be seen and heard. **Pintails** stick their tails high in the air, Shovelers cough and 'phuttt', Teal make a high-pitched 'bleat' that sounds a bit like a dripping tap, Wigeon make a delightful whistle and **Long-tailed Duck**, if you are lucky enough to hear one, can make an extraordinary honking, yodelling call. However, the most familiar duck, the Mallard can be seen displaying vigorously on village ponds the country over, even though birds will have paired up months ago.

Feeding fanatics

Another aspect that makes ducks irresistible is their feeding behaviour. For starters, what's a dabbling duck doing as it dabbles? Watch Teal, Wigeon, Shoveler and Mallard as they feed on the surface. They seem to be vibrating their bill. What they are really doing is acting like feathered filter pumps – sucking water in to the opening at the tip of the bill by a pumping action of the tongue. This water is then squeezed out of the sides of the beak through a set of fine grates (80-100 in a Mallard) leaving the goodies, algae, seeds and small animals, on the insides to be licked off by the tongue. They will often do this on the lea side of a body of water as any small particles will be concentrated by the wind in these areas.

All these birds have different-sized filter grates, with Shovelers being the real specialists, hence their wide bill. Look out for individual modifications of this technique by certain birds that will stir up water by paddling vigorously on the spot, up-ending to stir things up and even swimming backwards!

ABOVE: Goldeneye
When calling, Goldeneyes throw their heads back whilst making a creaking noise. This one, though, is asleep.

BELOW: White-fronted Goose (right) and Pink-footed Goose (left)
Wet pastures are the favourite haunts for the White-fronted Goose. The soft yodelling call as these birds fly in to their evening roost is another classic romantic winter sound.

Did you know ...?

It was after a visit by Sir Peter Scott to see the **White-fronted Geese** at Slimbridge in 1945 that he was prompted to found the Wildfowl and Wetlands Trust (for contact details see page 157). Over 4,000 Whitefronts still winter at Slimbridge, making it the most important site for this species in the UK.

RIGHT: Heronry
Grey Herons nesting are, if not among the earliest, certainly among the noisiest. Now is a good time to look out for lots of action at your local heronry.

Pterodactyls in the wood!

A few years ago I took some young birdwatchers to a local nature reserve. It was one of those days, and I had resigned myself to a Robin or two or a Nuthatch at best, when one of the more imaginative pointed skyward shouting 'Pterodactyl!' Now this would have been a scoop for Somerset. A large shadow drifted across the woodland floor accompanied by a variety of guttural grunts – a **Grey Heron**.

The heronry at this time of the year certainly sounds like a sound effect from Jurassic park with the breeding season for these early birds well underway. They use the same nests every year but still do a lot of spring cleaning and fine tuning and anyone watching a heronry

will see a steady stream of sticks being flown in. There is also a lot of posturing and noisy displaying as birds establish pair bonds and squabble over nests. The first eggs will be in place by the second week of the month and the action will then continue as different birds will lay at different times right through until September so there is plenty of time to catch up with your nearest colony. February and March for me are best because the action is not obscured by leaves.

Wild goose chase

Watching **Barnacle Geese** in Scotland – the first time you hear them it sounds like a million celestial Chihuahuas somewhere below the horizon. Strain your eyes in their direction and when they initially rise and break the line they resemble smoke particles. As they pass overhead, away from the coast to their feeding grounds out on the fields, they assume a well spaced order. With whistling wings tearing the air and adding to the cacophony, the whole spectacle of a flock of geese is so vast you cannot see all of it at once.

Now is a pretty good time to catch these birds, many of them winter migrants to our land. But you do not have to be in Scotland to get your senses around the seasonal goose. It is happening at wetlands, saltmarshes and estuaries all around Britain. There are plenty of them too – something like 480,000 belonging to ten species. Some of these you will have to be blessed to see, but it's always worth having a look because they do turn up from time to time. It's a bit of a laugh trying to work out which one of 30,000 silhouetted **Greylags** is a **Snow Goose** for example. It is also hard to miss and beat the fine harlequin plumage of a **Red-breasted Goose**.

More likely to be experienced in the South are the grunting flights of **Dark-bellied Brent Geese**, skeins of **White-fronted Geese**, with a **Pink-footed Goose** thrown in for good measure, and our most widespread goose, the Greylag. The ubiquitous introduced **Canada Goose** can be seen in its classic 'V' flocks, flying between ponds and reservoirs around our cities.

RIGHT: Greylag Goose
The original goose from which all domesticated varieties (except Chinese Goose) are descended.

Go for a Gander

ABOVE: Canada, Barnacle and Brent Geese
Although you may see Canada Geese on inland
waters throughout Britain, you may have to
travel to coastal areas to catch a glimpse of
wintering Barnacle or Brent Geese.

Caerlaverock, Dumfries (NY0365) – Imagine the sound of a single toy dog yapping then multiply this by 10,000, throw in the sound of 20,000 whistling wings and you are getting close to what it is like to be watching Barnacle Geese as they leave their nocturnal roosts on the Merse Saltmarsh and mud flats along the Solway Firth and head off into the fields to feed. There are now often Pink-footed Geese, Greylag, important numbers of Pintail Duck and the glamorous Long-tailed Duck to boot.

Loch of Strathbeg, Grampian (NK057586) – The list of web-footed winter residents to this fine RSPB reserve reads like a wildfowl hall of fame. This, the largest flooded dune slack in Britain and its shallow waters and associated freshwater marsh, can be home to over 30,000 birds, from our largest goose, the Greylag, to our smallest duck, the Teal!

Ribble Estuary, Martin Mere (SD428145) – A very important spot for Pink-footed Geese. Every winter over 10,000 check in from Greenland and Iceland. Some 75% of the world's Pinkfeet over-winter in Britain. Although most are in Scotland, the Lancashire coast and the Wash are the few places that you can see this goose in England.

Strangford Lough, Northern Ireland (468687) – There is plenty of opportunity for exploring this 29 km lough. It's the north end that is one of the country's best places to see Pale-bellied Brent Geese as they feed on the extensive Eelgrass beds. Other wonderful wildfowl that can be seen include Gadwall, Goldeneye and Mergansers.

Langstone Harbour & Farlington Marshes, Hampshire (SU687042) – Probably the best place in the UK for seeing Brent Geese. The 6,000 plus which regularly winter here are posers! And are very easy to get close views of from the sea wall that protects the marshes. This is also a good spot in the south for the dashing Long-tailed Duck and many other wildfowl and wader species. Black-necked Grebes can be a bonus.

Slimbridge, Gloucestershire (SO717048) – If wildfowl are your particular addiction, then Slimbridge is for you. Not only is there a huge collection of exotics and tame specimens of just about every British species but it is an attraction to really wild wildfowl too.

Amphibians, Reptiles and Fish

In early spring, the temptation comes to collect some **frog spawn** and watch one of the most mind-boggling processes in nature, from a blob of jelly to a fully formed froglet in a couple of months. But every year I get calls from people who find they are doing well and then suddenly all the tadpoles die.

RIGHT: Common Frog tadpoles one week after hatching
Every schoolchild's prize, Common Frog tadpoles are the product of the amphibious orgies their parents take part in every year.

To avoid many of the pitfalls, here are a few tips for good frogging.

1   Collect a small amount of spawn only (20 eggs for a small aquarium or tank is plenty).

2   Remember, as tadpoles grow their diet changes – they feed on algae to start with, then meat later on. Pond weed will supply the plant matter. Add in some fish food and tiny (1 cm cube) bits of beef on cotton will provide the meat. It is best to change the meat every couple of days.

3   Change the water regularly. Always use rainwater, not tap water.

4   Keep your tank in natural light rather than electric light, and avoid hot sunny windows.

5   As the tadpoles' rear legs start to form, reduce the level of water in the tank. Place plenty of moss, or sponge inside with the surface just above water level.

6   As the tadpoles start to leave the water, they must begin to eat insects. Either cover the tank with net and place it outside in your garden, or buy flightless fruitflies from a pet shop or reptile and amphibian specialist and place in the tank to provide food.

7   Once the tadpoles have turned into fully fledged froglets, release them back into the wild close to where the spawn came from in the first place.

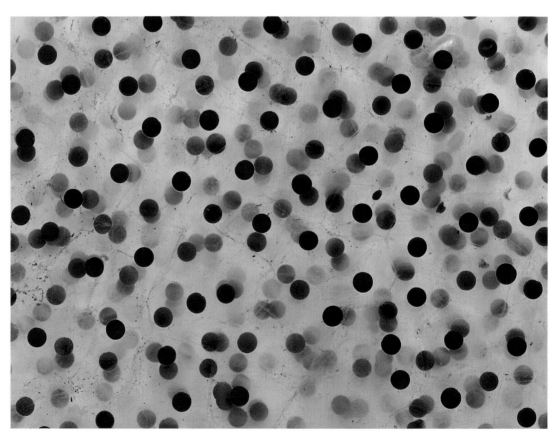

RIGHT: Frog spawn
The first sign of the fairer season approaching is the appearance of frog spawn. Say hello to spring's early manifestation! It's happening already in ponds in the south and west.

Invertebrates

For early insects spring is a thermal frontier. Their activity is limited by the temperature of the environment surrounding them, which is why it is only on these bright, warm spring-like days that you will see any appreciable numbers of insects at all this month.

Spring for an insect is an on and off thing, their behaviour is shackled to the ambient temperature. When it's warm their switches are flicked on, when it's cold they are off. But get a sunny spell and all sorts come out of hiding. As well as hordes of unidentifiable midges and flies, you will probably see the first of the season's butterflies rowing about the hedges and gardens looking for a few early flowers – these are species that hibernate as adults in sheltered, dry places. Most common are **Small Tortoiseshell**, **Peacock** and **Brimstone**.

These butterflies, particularly the first two, will spend considerable time basking, often utilizing bare open ground which heats up quickly. Once their body temperature is up to around 30° C normal activity can resume. Droneflies and various blowflies will also make an appearance in warm sun traps.

There are other insects that maximize any heat they find. Lift stones and paving slabs and you can find Black Pavement Ants' nests. The ants will be clustering around their queen, using the rock as a kind of storage heater – the rock warms up in the sun and releases this energy slowly to the ants below.

But the real thermal pioneer, able to put a buzz into even a cold day around this time are the **queen bumblebees**, of which there are several early season species to be seen flying this month. The reason that they can do this is that they are able to generate metabolic heat by relaxing and contracting their large thoracic flight muscles. They can do this, in a manner of speaking, by taking their wings out of gear and disabling the clutch-like attachment of the wing. Rather than the flying muscles working in antagonistic pairs, and almost oscillating within the insect's body under their own momentum, as they do in flight, they are fired by nerves to stimulate both pairs at the same time so any motion they might have generated is cancelled out.

This warm up occurs internally. A sign that something is going on is a pumping of the abdomen as the blood is circulated and a slight shivering of the bee's chassis. To help keep this heat in, the bumblebee also has thermal insulation in the form of its famously furry coat.

LEFT: Peacock Butterfly, BELOW: Small Tortoiseshell butterfly
The Peacock Butterfly and Small Tortoiseshell butterfly are both hibernators and among the first to greet the warmer days. Because they have been around since last September and still have a month or two to go, they are among the longest lived butterflies.

LEFT: Wood Ants waking up
Look for the winter-ravished nests of Wood Ants. Instead of impressive humps of frenzied activity, you will find a slowly seething tangle of ant bodies as they cluster together and use their combined dark mass to absorb as much of the weak solar energy as they can.

Plants

RIGHT: Alder catkins
Male Alder catkins are developing now, looking like purple pellets at the ends of the branches. Look for the distinctive female cones.

February's flowers are not necessarily the big and bold advertising campaigns of the summer months to lure insects, but they are flowers nonetheless and finding one or two of them can be as rewarding as a whole field later on in the year.

Many of our earliest native plants to flower do not depend on insects for pollination but depend instead on a much more reliable form of transportation at this time of the year, the wind. The most apparent are the male **Hazel catkins**, which in the more sheltered areas can be seen swinging languidly in the breeze. When they are at their ripest it takes a minimal disturbance to send a cloud of pollen spiralling off in the breeze. Blackbirds bursting through a well adorned hedge can look like they have been forcibly launched into the air by a well placed firework, accompanied by an explosion of custard-powder smoke. Much of this pollen is of course destined to waste, picking out the filigree of early spiders' webs or dusting damp branches, but with so much of the stuff in the air some finds its intended mark, a female catkin. Look closely at any Hazel branches and you may find the tiny bud-like structure with its tuft of crimson styles, looking like a strange arboreal sea anemone.

Other catkins at large now include probably the most splendid of them all: **Aspen** with its wizen-grey furry overcoat, which, when parted, reveals the deep purple stamens; the female catkins look similar but are tighter in structure. Look along rivers and waterways for **Alder**, another native tree, unmistakable owing to the presence of previous years' female cones. Look at the tips of the branches and you will notice the purple hue given to the branches is produced by many unopened male catkins.

The green shoots of **Bluebells**, like vegetational starfish, are beginning to peep through woodland floors along with the first sprigs of **Dog's Mercury**. At ground level things are happening too. It is often rather surprising to have your nostrils greeted by the out of place perfume of flowers this month. Many moments of searching for the source often commences. Is it coming from a garden, a tree even? Then notice, tucked away discreetly in the hedge, a collection

RIGHT: Cascading Hazel catkins
Just about every hedgerow has its subtle dusting of pollen produced by Hazel catkins. A breeze is all it takes to send a backlit hedge into a golden smoke as millions of genetic packets are launched on a passive journey. A few land on the weird, red female flowers and a few more make the nuts of autumn.

The red-purple female flowers emerge from buds in the centre of shoots.

Hazel has beautiful lemon-yellow male catkins.

Pollen grain

eyes will come to rest on introduced species of **Cedar**, **Cypress**, **Fir** and **Pine**. Bad news? Well in places they can be controversial; aesthetically not many people like to see the regimented order of a plantation. Within plantations and in the garden, ornamental varieties can grow so high and dense that they shade out the ground, smothering all that may have thrived below.

However, these same qualities can sometimes add to a landscape. In any big freeze, while the wind whips around the skeletal forms of our deciduous trees, the interior of a conifer wood is certainly warmer and provides more shelter.

My local plantation is a favourite playground at this time of the year – it is interesting in that it contains trees and vegetation of all ages and some clear-felled areas, providing a good range of habitat types and stages of growth. Get out of the car and you would be forgiven for thinking that it is a sterile place, with no sign of movement. But wait and listen and the audio clues become apparent – soft seeps, ticks, twitters and buzzes are often the first to be noticed emanating from up high. Keep watching and the makers will make themselves obvious.

There are plenty of small birds that pick away at pine seeds or insects that hide in the dense foliage such as **Coal Tits** and **Siskins** mixed in with a **Redpoll** or two. You may notice the leaf-like fluttering of our smallest birds, **Goldcrests**, and if you are not just lucky but also very observant you may make out among their number the brighter, bolder patterned **Firecrest**.

LEFT: Sweet Violet
Along with a small number of others, such as the petalled suns of Coltsfoot, these flowers are one of the few native plants around to seduce the earliest of insects such as Honey and bumble bees and Droneflies.

LEFT: Juniper
One of our native conifers, the Juniper grows up to 6 m tall and stands out this month as it is green all year around.

BELOW: Coal Tits
Coal Tits are among the commoner residents of the conifer plantation.

of rather odd-looking dull, mushroom-pink flower spikes that look like an alien interpretation of miniature conifers. These are the flower spikes of **Butterbur**.

Green stuff

Green stuff in the landscape is at a bit of a premium at the moment. Scan a winter scene and the dark green stains and smudges on hillsides within woods and gardens can belong to a number of species. There is **Holly**, **Ivy**, native **Scot's Pine**, **Yew** and **Juniper**, the last three being our only native conifers. But more than likely your

March

March for me is when spring has undoubtedly started – look closely in every habitat and you should find little fragments of optimism, even if the weather conditions say otherwise. The birds that wintered here are in the full swing of the breeding season, with territories well established – the migrants are well on their way. Flowers begin to brighten our days and insects tentatively emerge.

LEFT: Brown Hares
The 'mad' behaviour so often quoted is not male hares doing battle, but a female trying to put off an over-amorous suitor.

Mammals

The key mammal to look out for in March is the **Brown Hare**. Even though it's a spring cliché, March is the best time to see them. Hares can be seen courting in any month, but they could be obscured by lush spring growth later in the year.

Hares occur all over the UK except Ireland – I find hot spots to be East Anglia, Wiltshire and Hampshire. A train journey passing through lowland pasture and ploughed fields can be a good way to locate hare populations. When you know where your hares are, it's simply a case of using good field craft. Wear muted colours, keep quiet and move slowly keeping the wind in your face and you should be well rewarded.

Hares are not as common as they used to be. Modern cereal farms provide little or no food in autumn, and livestock farms have few crops for them to hide in.

Adult females lack the distinctive black mask and are usually a duller grey or brown on the upperparts.

Adult males have a grey back, black wings and mask.

Birds

Many people have a particular event above all others that hails the coming of warmer months. The reason I can confidently state that spring is here, is that I can bet that on visiting my local patch of moor I will find, waiting patiently on a granite boulder, amid a desolate and often unwelcoming habitat, a male **Wheatear**.

These birds will have undergone a migration, some 10,400 km (6,465 miles) to the UK, from wintering grounds south of the Sahara in West Africa in order to breed on our mountains and moors. Anyone can catch a glimpse of this bold bird, one of the smallest long-distance migrants, by intercepting them on landfall as they cross from the Continent to our shores. Birds arrive in the south-west first but can be seen anywhere along our southern shores towards the end of March, peaking around the first week of April.

Following the first wave of migrations there is a second peak as the bigger, leggy, longer-winged birds pass through, such as **Grey Plovers** and **Curlew Sandpipers**, destined for Iceland. These birds do not breed in Britain.

Stimulated by slowly rising temperatures and longer day length, many birds will start to display and strike up the first bars of the dawn chorus. **Lapwings**, the commonest British plover, will be setting up territory. The males perform their noisy tumbling flights, looking like they have suffered a mid-flight coronary attack and come crashing down to the ground only to recover at the last moment with perfect gymnastic poise.

Knock on wood

Great Spotted Woodpeckers normally spend a lot of their time hidden from view and, unless you are lucky enough to have one visiting the bird table in your garden, you will probably not have seen one up close.

Choose a warm day and head for the woods. Take a dead stick and find a hollow-sounding tree trunk and hammer in short bursts as fast as you can! Woodpeckers do this instead of singing like most birds. With luck, your woodpecker impression will be good enough to attract a territorial male that will think he has a rival to chase off his patch. He will check you out before realizing he's made a mistake. If you are lucky, your banging may attract other birds such as Nuthatches and even the **Lesser Spotted Woodpecker**.

Feather a nest

You've heard of putting food out for the birds during the winter, but have you ever tried putting nesting materials out in spring? Collect sheep wool from barbed wire fences, horses' hair, feathers and even dry grass and hay and the fluff from your Hoover bag. Bundle it up and hang it around the garden or on the bird table

ABOVE LEFT: Long-tailed Tit nest,
ABOVE RIGHT: Robin singing
It is time to establish resources for the coming breeding season and, even if the weather has been barely warm enough for you to emerge,

you can bet that the natural world has been busy. Most obvious are the birds. Robins are singing their larynxes out and many birds, such as this Long-tailed Tit, will have started work on their masterpiece nest.

and see what comes and takes what for their nests. If you watch which direction the birds fly with the building materials you may even be able to work out where their nests are.

Learn the chorus

Now is the best time to start learning the bird's chorus. With just a few birds singing at a time, you can become familiar with the main players before the avian world really starts swinging and Britain is swamped by the complicated songs of the various warblers.

This is something that anyone can do, whether they live in the town, city or country. In fact the urban situation is often easier as with less cover many birds utilize fairly open manmade perches such as telegraph poles, pylons and aerials for their own form of communication. It certainly adds to the experience if you can reveal more than simply which species makes which noise.

RIGHT: Common Buzzard
Look out for Common Buzzards in open country soaring high in the sky, with wings held in a shallow V-shape.

While learning the songs of the more familiar such as **Robins**, **Blackbirds**, **Song** and **Mistle Thrushes** you will quickly understand one of the more important roles of singing and that is in the staking out and establishment of territories. By watching the birds and plotting on a map their favourite singing perches, which are often central and overlook as much of the territory as possible, and by watching the birds as they fly between perches or have scuffles with neighbours, it will only be a matter of time before you really get to know the size and limits of the territory of the birds on your patch.

Birds of prey

Normally the laziest of birds of prey, **Common Buzzards** can also be spurred into uncharacteristically active swooping displays. Look out for them in mixed woodland, near farmland and also on moorland and heaths.

Look out for the undulating, swooping dives of **Sparrowhawks** as they set up territory over woodlands.

Amphibians, Reptiles and Fish

The water bounces the blazing early morning March sky back at me. It seems to defy gravity and bulge upwards in the middle, with an odd kind of lumpy texture. Every few seconds the whole reflection wobbles and shivers as something moves in the rushes at the edge – and straining my ears I can just about make out a

bizarre purring noise, like a tiny electrical generator – the frogs are back!

Common Frogs are the earliest of our amphibians to emerge from hibernation and in exceptional years have been recorded breeding as early as January in the warmer south-west. Like all amphibians, on emerging from hibernation they head straight for their breeding ponds,

BELOW: Common Toads mating
Toads tend to spawn later than frogs. Toad courtship is not romantic; there can be so many males trying to mate with a single female that females occasionally drown.

unless of course they are one of the males that has adopted the strategy of sitting it out in the debris or mud at the bottom of the pond. This they are able to do as they can breathe sufficient quantities of oxygen from the water through their skin. (These males might get the returning females first but in a harsh winter risk being frozen solid if the pool is a shallow one.)

The pre-spawning build up is a gradual and unspectacular affair as frogs slowly accumulate, with most of their movements occurring under the cover of darkness. Nip out with a torch on a wet, warm night when the ambient temperature ranges from 7–10° C and you are likely to see frogs bobbing around in the pond and scattered around the surrounding habitat.

If they haven't started spawning yet, keep checking up on them, because at some mysterious cue the pond will erupt into an amphibious orgy of splashing, spawning frogs. This 24-hour a day spectacle will last between 1–5 days. Approach the pond slowly, keeping low and quiet and establish yourself in a comfortable position (warning – comfy positions are few and far between around ponds). All those individuals that rumbled your approach and ducked below the surface, stopped calling or froze will soon get distracted by their surging hormones and when a few minutes have passed you will be enjoying a frog frenzy.

LEFT: Common Frog, spawn and tadpole
Common Frog tadpoles will be getting quite large in size in the warm west, but farther north and east they may yet be twinkles in their mothers' eyes!

BELOW: Common Toad, spawn and tadpole
Toad migrations to their breeding ponds are often rather large and visual affairs. Probably the best way to witness this is to contact your local Wildlife Trust and offer your services in one of the many nocturnal toad road crossing schemes that are going on throughout the country.

RIGHT: Smooth Newt
The commonest of the newts, Smooth Newts resemble a slicker and smaller version of the Great Crested Newt. The main differences between them, apart from the smaller size, are the very smooth skin, with no warts, and a much lighter skin, with more black blotches down the sides.

Things to look out for are: the subtle difference between males and females, the former are generally darker and smaller than the females with a bluish tinge to their throat and in some populations the females often look reddish. Very obvious also will be the mating hug. Technically referred to as 'amplexus' this is when the males climb aboard a female and hang on tenaciously with their front limbs, helped by rough black swellings on their thumbs known as nuptial pads. Assuming the males don't get displaced by larger rivals they will hang on until the female

is ready to spawn (usually at night) and as soon as she expels her eggs, the male will shed sperm on them as they leave her body.

It's not just Frogs that are jellying up our ponds, dykes and ditches. Five of our six native amphibians (Common Frogs, **Common Toads** and the three species of **newt**) can be found breeding this month. Reptiles such as **Adders** will begin emerging from hibernation.

Fish linger

All it takes is for the moon, earth and sun to be aligned to make a very happy rock pooler this month. Around the spring equinox are the largest tidal ranges of the year, exposing the very lowest regions of the shore, a realm normally out of reach unless you are prepared to go snorkelling. As you follow the receding tide, it is worth exploring among the rocks and boulders.

March is the best month to find adult **Lumpsuckers**. This fish is about 40-50 cm long, colourful and rather comical, looking a bit like a knobbly deflated balloon. In March the males and females come inshore to breed but rarely make it into the intertidal zone, preferring to lay among the kelp out of reach of the normal tide cycles. The female then departs for deeper water leaving the male to guard the eggs and ventilate them. So dedicated is he to his task that this month's spring tides often catch him out. These ponderous and unlikely looking fish are often stranded in low-tide pools alongside the 30,000 odd eggs left in their charge.

LEFT: Lumpsucker
The Lumpsucker has the ability to adhere to rocks using a sucker created by its pelvic fins. This, coupled with the male's often striking bright pinky-red belly, makes it one of Britain's most spectacular and oddest fish.

Invertebrates

Spring is a notoriously unpredictable thing! Most species take their cues from a combination of air temperature and day length. The small rise in temperature that accompanies sunlight acts as a trigger.

The air temperature may be cold but sheltered sunny spots such as sun-warmed walls, trees and hedges are the sort of place that can be buzzing with the sound of tiny cellophane-thin wings. Listen in the spring air as you blunder close enough to disturb sunbathing blowflies and hoverflies lured out of suspended animation.

Butterflies may be seduced by spring sunshine. Species to look out for include **Small Tortoiseshell**, **Peacock Butterfly** and the original butter-coloured fly the **Brimstone Butterfly**.

Tiny caterpillars of some of our scarcer butterflies are crawling out of hibernation – **Marsh Fritillaries** have been walled up in their own communal survival chamber of silk; they will emerge and bask in the sun clustered together. Their black bodies warm quickly and, once up to temperature, they begin the second leg of their lives as caterpillars. **Silver Washed Fritillary** larvae take it a step further: they emerged from the egg late last summer and, without even eating a bite, they hid in a crevice or the moss on

the bark of the tree where they were laid back in August – only now will they crawl down and seek the sustenance of Violet leaves.

A penchant for pollen (and nectar)

Believe it or not there is a word to describe standing around at the base of a tree at night. It is known by entomologists as 'Sallowing'. Not just any night will do – the warmer the better – and not just any tree – it should be a flowering **Sallow** tree. To complete the act you need a torch, oodles of curiosity and luck.

LEFT: Sallow catkins
Sallow catkins are the big nectar source at the moment. Unlike other trees that have catkins, Sallow relies on insects, rather than wind, to distribute pollen, hence the big buzz around these trees this month.

BELOW: March Moth
Its name may not be as poetic as a lot of other moth species, but you cannot argue with it. The males (females are flightless) can be found on the wing from March through to April.

LEFT: Marsh Fritillary caterpillars
On those same bright, crisp, first days of spring when we finally feel like crawling outside to soak up some sun, so do the caterpillars of the endangered Marsh Fritillary.

RIGHT: Buff-tailed
Bumble Bee (Bombus
terrestris)
This is not only one of
the first bumble bees on
the wing in spring, it is
also the largest of our
18 species.

RIGHT: Comma
butterfly
Comma butterflies are
common, but absent
from Ireland and
northern Britain. It
flies from March to
early September.

FAR RIGHT: Meadow
Bumble Bee (Bombus
pratorum)
Appears in early
spring, visiting both
long, tubular flowers
and open flowers. Nests
on, below or above
ground, including in
nestboxes.

can be found embedded in the fluff
as they strive to gorge on the plentiful supplies of pollen and nectar.

Spring skittles

If you've tried to get a spring fix and
are failing then a trip to a rocky shore
should put that right. Look under
craggy overhangs for the tiny skittle-
shaped eggs of **Common Dog Whelk**,
a pale swirly shelled sea snail.

In the same sort of habitat but not
quite so conspicuous are the flattened urn-shaped version turned out
by the close relative the **Netted Dog
Whelk**. Look out for other reproductive curios that signal the oncoming
of spring. Green blobs of jelly are the
eggs of the **Green Leaf Worm**. Pink
string wrapped around weed and
rocks are the egg masses of a sea slug
called a **Sea Hare**, while white frilly garters
found discarded on the shore are the egg rib-
bons of **Sea Lemons**, not a fruit at all but
another sea slug.

Some of the rocks will appear fluffy, with
new growth of the more brittle seaweeds
destroyed by the scouring winter waves, and
these in turn will be grazed by the ever-present
herds of molluscs. Many species can be found
scattered over the rocks like large living hundreds and thousands. The dark and duller beasts

Get it right and your torch will pick out the
shining eyes of many of the early nectar-feeding moths, such as the Noctuids: **Hebrew
Character** *Orthosia gothica*, **Clouded Drab**
O. incerta and **Common Quaker** *O. stabilis*.
These moths in turn become the focus of early
bat activity.

If, however, you are not much into late
nights, Sallows are just as rewarding during the
day. In fact when the rest of the countryside
seems devoid of insect life, the Sallow is the
place to be as it seems like a rendezvous for
everything with wings and a proboscis. Backlit
against a low sun the furry haloes of the catkins
are interrupted by the dark giant silhouettes of
Peacock, Small Tortoiseshell and **Comma** butterflies. While more subtle, early **hoverflies**, such as
Syrphus vitripennis, *Eristalis intricarius*, **Honey
Bees** and the first of the **queen bumble bees**
like those of *Bombus terrestris* and *B. pratorum*

RIGHT: Hoverfly
(Syrphus ribesii)
Another early flier,
males perch on leaves
or twigs up to 2.5 m
from the ground and
make a high-pitched
whining noise.

ABOVE: Common Dog Whelk
For the spring-time optimist the rocky shore is the place to go. The Common Dog Whelk is laying eggs and has been since February.

ABOVE: Sea Hare
Looking like a Magic Roundabout extra, the Sea Hare is a surprising mollusc, especially when you meet a big one; it grows up to 20 cm!

are **Common** or **Edible Periwinkles** *Littorina littorea* and among them are their more numerous and flamboyant brethren the **Flat Periwinkles** *Littorina mariae* and *L. obtusa*, easily separated from other colourful species like the **Rough Periwinkle** *Littorina saxatilis* by their flattened coiled spire.

Look out for the first generation of worker bumble bees – the little bees flying around. The reason they are so small is that they were reared single handedly by the queen on a small amount of food, so they turn out a little stunted.

Bee box

You may see queen bumble bees out and around the garden at the moment. They are beginning to look for a place to nest, and since they are good for the garden how about making a nest box for your very own bumble bee colony? There are no rules as long as the bees can get in and out, there is ventilation and nest material, and mice and other predators cannot get in (see box).

Practical bit

Bee box
Take a large flower pot, turn it upside down, cut four holes in the rim of the pot so these make entrance holes. Make an umbrella (to cover the top of the nest, and stop rain falling down the hole) out of the bottom of an old plastic drinks bottle. Glue this in place.

A small piece of chicken wire rolled up into a basket filled with moss, grass and cotton wool can then be scrunched up inside to form the nest. Then you just need to make a base. Take a large flower pot drip tray, place a cork in the middle (this is removed later to form a drainage hole) and pour plaster of Paris into the tray and leave it to set.

Remove the cork, lift out the plaster base and glue the top to the base using bathroom sealant. Cover the whole thing with grass and leaves, making sure the entrances are exposed and wait for your bees to colonize.

LEFT: Bombus hortorum
Like many insects, Bombus hortorum is known best by its scientific name, but, having said that, it's quite a user-friendly one. Three yellow bands and a white tail identify this bumble bee.

Plants

As a knock-kneed schoolboy I would notice that on the way to school before the sun had touched their petals, there would be nothing and on the way home that same day, the same path would be lined with golden flowers. When I realized the flowers opened and closed on a daily basis, I remember sitting by a vibrant clump of **Lesser Celandines** and waiting for the moment when they shut up shop. I never did witness it – firstly it isn't as instant as I wished and when you are small your attention is short and easily distracted.

Nowadays other than simply appreciating what was Wordsworth's favourite flower I tend to spend my time crawling around looking at the leaf junctions with the stem. If I find tiny bulbils nestling here it indicates a subspecies known as *Ranunculus ficaria bulbilifer*, a plant which has sterile flowers but relies on trampling animals, wellies and floods to disperse these tiny units of propagation.

The first rose

Not a rose at all, **Primroses** get their name from the words *Prima rose*, meaning the first rose. There is more to their flowers than simply being pretty splashes of floral gold in our gardens and hedgerows.

ABOVE: Primroses
Spring gold - a bank of Primroses is a soul-warming sight. Many of the early insects, such as Brimstone Butterflies, hoverflies and beeflies, depend on them for nectar.

RIGHT: Lesser Celandine
The reality of the Celandine's magical disappearing act is down to the sepals underneath the petals being green, so that when the flowers close the gold dissolves into the surrounding foliage - an adaptation to protect the flower interiors from rain and frost.

Daffodil Days

Lea and Paget's Woods, nr Hereford (SO597342 & 598344) – Two blocks of woodland that offer just about all the atmosphere you could ever want from a wood in spring. Take in the good show of wild daffodils with the other classically vernal sounds of a Chiffchaff going off in the Hazel thickets and the territorial drummings of Great and Lesser Spotted Woodpeckers.

Wentwood Forest, Gwent (ST427949) – Mixed woodland, mainly larch, doesn't fit the classic wild daffodil scene. However, there are patches of old pasture present here and the daffodils look utterly gorgeous!

Dunsford Wood, Devon (SX792885) – One of my favourite haunts, it has the best display of wild daffodils that I have ever seen. On top of this it has a great warm Devon river valley atmosphere. Although it is still a little early, whilst enjoying the daffs look out for the large metallic glossy blue-black bodies of Oil Beetles, which are often found nibbling their leaves.

Farndale, Yorkshire (NZ662001) – A very famous site for 'Lent lilies'. The flowers occur in abundance among the valley meadows that the river Dove has carved out of the North York Moors. Whilst in the area it is worth nipping up to the moor to listen out for the rumblings, growlings and repeated 'go back' calls of Red Grouse.

The 'Daffodil Way', Gloucestershire/Herefordshire (SO701312) – This is ten miles (16 km) of spring-time WOW! A circular route through some of the best daffodil locations in the country. For a leaflet ring Newent tourist office on 01531 822468. Also nearby are two of the Gloucestershire Wildlife Yrust's reserves, Gwen & Vera's fields (SO696277) and Betty Daws' wood (SO696284) contact the trust for further details.

West Dean Woods, nr Chichester, Sussex (SU849160) – One of the best locations in the south-east for wild Daffodils, a former stronghold. Worth the trip to see the big March yellowing beneath the coppiced Hazel. This wood is mostly private although there is public access at the above grid reference. The Sussex Wildlife Trust, who manage the reserve, holds open days at the end of March. Tel: 01273 492630 for more details.

ABOVE: Wild Daffodils Simple, small and perfect and about as far as you can get from the gaudy mutants produced by the horticultural world.

Pin- and Thrum-eyed Primroses

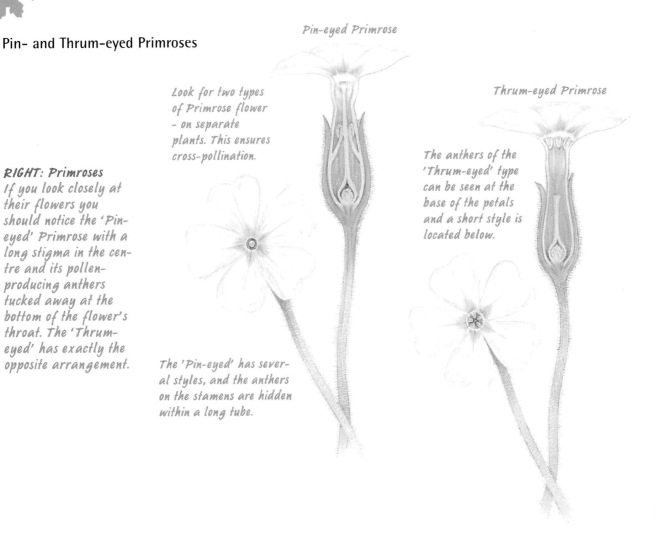

Pin-eyed Primrose

Thrum-eyed Primrose

Look for two types of Primrose flower – on separate plants. This ensures cross-pollination.

The anthers of the 'Thrum-eyed' type can be seen at the base of the petals and a short style is located below.

RIGHT: Primroses
If you look closely at their flowers you should notice the 'Pin-eyed' Primrose with a long stigma in the centre and its pollen-producing anthers tucked away at the bottom of the flower's throat. The 'Thrum-eyed' has exactly the opposite arrangement.

The 'Pin-eyed' has several styles, and the anthers on the stamens are hidden within a long tube.

Flower diversity is still not at its height and those that are in bloom, like our Primrose, tend to do so in high quantity – a resource tapped by the early insects. There is an inherent problem here. The whole point of flowering is to spread the genes of an individual via male pollen grains to the female parts of another flower – this is plant sex. The danger is with so few species in flower and many producing more than one flower per plant there is a risk of self-pollination and that would defeat the object of flowering. Primroses get over this by a nifty arrangement when it comes to the male parts (the anthers) and the female parts (the stigma and style). There are two forms that differ in structure.

This means that an insect visiting one plant and one kind of flower will pick up pollen on a certain part of its anatomy and the pollen will not get transferred to the female stigma of flowers of the same type on the same plant. If the insect visits a flower of the opposite arrangement the pollen will be in perfect alignment – self-fertilization is avoided and the desired exchange of genetic material is achieved!

Bawdy botany

No flower can be ruder than the **Arum Lily** *Arum maculatum*. For a start it has a list of traditional names containing more innuendo than a *Carry on* film. These are of course inspired by the famously phallic spadix, a purple spike that is sheathed by the hood and produces a putrid smell guaranteed to attract a desperate fly or two. Names include Lords and Ladies, Vicar in the Pulpit, Willy Lilly (a favourite of mine) and Cuckoo Pint (the latter word is slang for penis and if you follow the origin of cuckoo, through cuckold, back to its roots this can mean erect!).

As well as wafting its stench to attract insects, the flower is aided by a metabolic heat produced by the spadix as well – this flower looks interesting through a thermal imaging camera! Insects attracted crawl down past a hairy valve of backward-pointing hairs and find themselves trapped – bouncing around between the male and female flowers they self-pollinate the plant – escaping when the flower withers. Later on the female parts will develop and swell, replacing the spadix and spath with a turret of scarlet berries.

Arum Lily – how it works

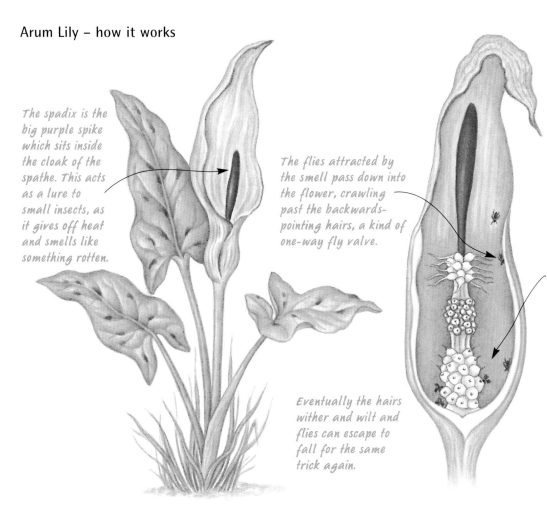

The spadix is the big purple spike which sits inside the cloak of the spathe. This acts as a lure to small insects, as it gives off heat and smells like something rotten.

The flies attracted by the smell pass down into the flower, crawling past the backwards-pointing hairs, a kind of one-way fly valve.

Eventually the hairs wither and wilt and flies can escape to fall for the same trick again.

LEFT: Arum Lily It's a trap! The weird and wonderfully stinky Arum Lily. It's a hedgerow spectacle in the spring and one of our most interesting plants.

Once trapped, flies have to earn their release by pollination. They ensure this by moving between the male cluster at the top and the larger female flowers at the bottom.

The female flowers, when fertilized, become those equally intriguing clusters of orange and scarlet berries that you'll notice later in the summer.

Fallen stars

One of my favourite scenes is the delicate phosphorescence that the waking of a thousand **Wood Anemones** lends to the floors of some of our more open woodlands.

Depending on conditions, they can also animate the landscape as they shiver and nod, pushed by the March winds channelled through clearings and rides – a sight that may well have led to their other old name of wind flower.

On a still day they play with another sense. Without the breeze their weak perfume is allowed to concentrate and a walk among them will often reveal pockets of scent, a sweet musky smell.

This soft-flowered plant is a clear indicator of ancient woodlands. In some places they are the ghost of landscapes past as the flowers linger hundreds of years after the original wood was cleared. The reason they are loyal to their position is that they rarely set seed and so rely on vegetative spread by runner and sucker, spreading at a rate of around 2 m every 100 years or so.

LEFT: Wood Anemones Like fallen constellations, Wood Anemones track with the sun and are really one of the most delicate and subtle of the floral tributes to spring.

What is spring? A magical rebirth, a time of renewal – perhaps it is summed up in one word ... sex! Freed from the thermal, physical or physiological constraints of winter, the natural world makes up for lost time and individual species get on with the business of spreading their genes around. The business of procreating can be witnessed by anyone everywhere they care to look, listen or sniff. Spring can be as much the whiff of a freshly trodden bed of Wild Garlic, or the lethargic trundlings of an Oil Beetle, as it is the mass flowering of Bluebells or the first evocative calls of a Cuckoo.

Mammals

It's a bit of a spring wildlife cliché. It is so entrancing and fantastic that it seems it could only be an image from a wildlife calendar. But **Badgers** really do frolic in the **Bluebells**! And April is the month for both Bluebells and copious amounts of frolicking. So if you have never sat in the waning half light of a wood in spring waiting for Badgers to emerge from the dusty depths of a sett, now is definitely the best month to do it, for several good reasons.

LEFT: Wood Mouse
The Wood Mouse is one of our more common wild rodents. Its big eyes and ears, and long tail and legs make it very agile - perfect for the complexities of woodlands.

One good reason, the dawn chorus, is well known and peaks towards the end of April, but very few people actually make that early 5am start. However, because you need to be in position a good hour before sunset by the sett you want to watch, you are a captive audience for the reprise, the dusk chorus. As sight dulls with the fading light, sound takes over. This is the same for the many birds that carry out most of their activities by sight, so as the light fades the only useful thing they can do is sing and re-confirm their hold on a particular territory. Listening out for that last burst of song at the handover of day to night, followed by a moment's silence, when watches tick and you can hear your heart beating, is something very special. A **Wood Mouse** rustles by your feet and, abruptly, the silence is shattered and night relaxes into its own folds.

Brock watching

Another good reason is that this month sees a riot of activity around the sett. Badger world is very much alive, even though that first moment, waiting for the humbug-striped head to crane elegantly and cautiously out of the pitch, could almost be any season. After the emergence drill; nosing the air, retreating, sniffing the air again, followed by a long bristly scratch, the scene changes into a very different one, one that can only be April. The cubs emerge this month, having been underground since around February. On top of all this there is plenty of other activity, with digging and the collection of bedding, which is often hauled long distances.

The Badger cubs make up for a cramped two months underground, careering around like fluffy bumper cars. They can get so carried away with themselves that, on a couple of occasions, I have stood in the shadow of a tree bole having very cautiously followed the rules of Badger watching, and been rumbled by a feisty cub running straight into my legs.

There really are few more accessible and predictable animals than Badgers. Find a bit of woodland on a hill, whether in the country or on the outskirts of town, and more often than not they'll be living there.

It has always struck me as being rather sad that the Badger, probably one of the most endearing and charismatic animals in Britain and also one of the commonest, is seen alive and in context like this by so few. More often than not they are only noticed in passing as a crumpled and broken heap on the hard shoulder – one of the 37,000 or so victims to our road systems every year. Badger watching is often portrayed as an activity for iron-willed field naturalists, and this may seem to alienate the rest of the population. So if the thought of creeping about after dark, getting cold, wet and mosquito-bitten just isn't you, don't dismiss the idea of ever being able to witness the sorts of scenes described earlier. All over the country there are organized Badger watches, where wild Badgers can be observed relatively

What to look out for this month

- Baby Badgers
- Molehills
- Cuckoos
- Newts
- Slow-worms
- Orange-tip butterflies
- Bluebells

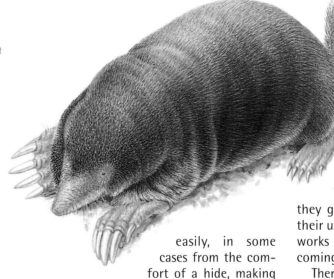

RIGHT: Mole
Anyone who has a lawn
will probably notice
that Mole activity
increases as the soil
warms up. This is due
to worms heading into
shallower soil and the
fact that Moles are
breeding now.

easily, in some cases from the comfort of a hide, making the whole experience open to everyone. Then it's just a matter of time before the addiction gets you and soon you too will be regularly sitting out among the Bluebells!

How to get them to linger? Unsalted nuts and bread are fine but the Badgers hoover them up quickly. If you scatter too early the birds will eat them; too late and too close to dusk and you risk disturbing the very mammals you set out to watch – Badgers can be extremely sensitive to disturbance.

Here is a trick that seems to prolong the moment of Badger-watching bliss, but bear in mind if someone sees you doing this it can be a little embarrassing. The trick is to visit your sett during daylight and spread unsalted peanut butter and/or honey on tree stumps, bark, logs and even on the leaves. Badgers have a real weakness for these two foods and when they emerge they will busily lick it all up.

You get to see Badgers for longer and they get a snack that must taste better than their usual food of worms and beetles! This also works well in the garden if you have Badgers coming to visit.

There are a variety of Badger Watches around the country. A few are shown on page 59.

Did you know . . . ?

Go out **Mole** watching and if you see signs of activity and notice one molehill that is larger than the rest, this may be a fortress, the molehill that contains the football-sized nest in which the young are reared.

RIGHT: Feeding
Badgers
Badgers are one of our
most familiar and
common large animals,
but despite this, very
few people have actu-
ally seen one alive. Try
to make sure you are
among the exceptions!
Badger Watch schemes
provide wonderful
opportunities to see
these shy mammals.

Birds

Birds are the soundtrack of spring. They are at their most vocal now with the breeding season well underway. Males in particular are trying to 'out flash' their neighbours. **Starlings** sit high on roof tops, whistling, wheezing and clicking their way through a repertoire, whilst **Blackbirds** and **Song Thrushes** will regularly gush forth their phrases from a vantage point in their territory. Many of these birds are originally woodland species and rely on song to advertise their territory occupancy and health to neighbours and potential mates and, in habitats that can be dark and visually obstructive, a good song makes sense. But if you have a clear air space with good visibility then you can add another dimension to your genetic advertising campaign – stunts!

Birds of open places are the undoubted masters of aerobatics. **Skylarks** perform a Red Baron style rise and descent to several hundred metres sustaining their warbling as they go, falling silent as they approach the ground. The closely related **Woodlark** in the south does a song flight that circles around whilst singing. Many phrases are uttered but listen for the 'lu-lu-lu' bit, this is where 'Lullula' comes from in the bird's scientific name, *Lullula arborea*.

ABOVE: Cuckoo
That traditional harbinger of spring, the Cuckoo's call is very familiar. The female Cuckoo removes an egg from the host's nest before laying an egg of her own.

ABOVE: Skylark
Skylarks proclaim their territories from on high, a seemingly endless, trilling warble.

RIGHT: Starling singing
The Starling pours forth its mix into the dawn chorus, mimicking many other birds' songs plus making up bits as it goes along.

My favourite bird of April has to be the **Tree Pipit**. The males arrive from Africa this month. When displaying it rises vertically from a tree or bush and then on the descent it fans both tail and wings and parachutes slowly back to its song perch singing a far-carrying melodious warble as it goes. It can be separated from the commoner **Meadow Pipit** by the fact it starts its singing from a tree and not the ground and also ends with a three-note 'zee-zee-zee' trill.

Egg and chips

Bird egg shells, chips and fragments are often found on lawns and pavements. They may have hatched naturally or been predated. Those that have hatched have been opened neatly by the chick pecking a line around the shell and popping off the cap, the shell will be neatly broken with the edges opening outwards. The biggest clue is if there is an intact membrane. Predators such as birds and mammals make a messy job and often leave punctures and cracks, and normally clean away most of the membrane. The presence of egg shell doesn't mean there is a nest nearby. Often they are removed and dumped by the adult bird far away so as not to attract the attentive eye of predators. A complete egg, often pale blue, can be explained by the behaviour of parasitic Starlings, which often nip into a neighbour's nest, lay an egg of their own and remove one of the original clutch.

Pollen pirates

With attention drawn towards the first of our garden flowers at the moment, it is worth keeping an eye on the birds too – namely that common garden resident the **Blue Tit**.

This bright (in more ways than one) bird in recent years has been reported sticking its beak in places where it previously hadn't been recorded. Namely within the flowers of various 'pussy willow' catkins and the nectar-rich flowers of garden classics such as those of Mahonia. Small birds like the tits are very active, holding territories and chatting up females, mating and squabbling. This is a rather unpredictably sparse season for food. So what better food supply than the energy and protein-rich pollen and nectar provided by plants for a chiefly insect clientele?

The benefits of this relationship are twofold. It has been estimated that if a bird spends three hours with its head immersed in flowers it can gain the equivalent of a whole day foraging for seeds and insects, leaving plenty of disposable energy for breeding activities, and the plants, it is assumed, get another agent to distribute their pollen for them, especially useful if the insects are a little thin on the ground after a cold snap.

Amphibians, Reptiles and Fish

Investigate a weedy pond at night and in your torch light you may find any one of our three British species of newts, **Smooth**, **Palmate** and the rare **Great Crested**. They have returned to water in the last month and like their more vocal and well-known relatives the frogs they are also here to mate and spawn.

What they lack in noise, they make up for in poise, posture and colour. The males in particular are splendidly bedecked in their finery, with membranous scalloped crests running the length of their body. The fingers of spring turn up both the contrast and the colour, and flanks of all species become a rich collage of orange, blue, white and black. If you are lucky you may witness the courtship dances in shallow water as the male does a flamenco dance. Fanning his tail around the female's face, he leads her quite literally on a merry dance. His choreography has to be precise as he deposits a capsule of sperm and directs her into the correct position to pick it up.

Slug guzzlers

A clear sky and a sun trap hedge make for perfect reptile watching. Start early in the day and work your way around field edges and waste ground and you can turn up the first reptiles to emerge and soak up the sun, **Grass Snakes**, **Common Lizards** and even the mass emergence of **Adders**, which often hibernate communally.

One that may elude your eyeballs is our other common lizard, the one without legs, the **Slow-worm**. It does bask, but rarely in full view. I have some in my garden and the most I see of them is a tail sticking out of a hole or a golden flank just visible through a tangle of vegetation, and even then only when I have got my eye in.

At this time of the year I tend to forego my gardening intent and wander about my patch doing a census, lifting up bits of carpet and corrugated metal sheets that I leave lying around as here the Slow-worms tend to congregate, whether post hibernation or ready for mating I'm never sure. All I know is that *now* is the time to see them, to look for the females with the dark stripe down their sides

ABOVE AND RIGHT: Newts mating
Look into the shallows of the pond and you may see newts mating. The male, with his flouncy frills and bold marking, woos the female with a tail-flicking performance. You are less likely to see the female laying her eggs. The act looks like she is making a snowball, using her rear feet to wrap a living leaf around each egg, but if you search hard enough you will find leaves bent over, containing a single egg.

and the males looking like long lost golden amulets bejewelled with blue flecks about the head indicating that they are in breeding condition. By June, no matter how hard I look, I cannot find a single individual.

BELOW: Slow-worm
Eyeball to eyeball confrontation! The Slow-worm is a lizard without legs.

ABOVE: Orange-tip butterfly's eggs
You do not have to be a butterfly expert to find the eggs of Orange-tip butterflies. They lay on few plants in rather obvious places, on the flowers and buds. The eggs turn as bright an orange as the males' wing-tips. What could be easier?

Invertebrates

It's not just the furred and the feathered that go into overdrive. It takes a warm April day to make you realize how much you missed insects.

The buzz is back

With a well choreographed synchrony they make a mass return to our attentions, as numerous species come out of their suspended animation as pupae, eggs or hibernating adults and reconfirm their six-legged grip on the countryside.

Probably butterfly of the month has got to be the **Orange-tip** or Wood Lady, the male of which fits its common name rather well, and cannot be mistaken for any other British butterfly. Orange-tips time their one annual flight period with the opening of the flowers of their caterpillars' foodplants, various members of the Crucifer family, in particular the pale purple-pink of Lady's-smock flowers, a plant that can equal the Bluebell spectacle in certain damp unspoiled grasslands.

These butterflies have little to fear from predatory birds; as caterpillars they sequester the mustard oils of their food plants. The females, which lack the orange, have even less to fear as they spend more time lurking.

Also boldly going where no pleasant-tasting insect would dare, **Bloody-nosed Beetles** are creeping around in the warmest corners of the south, they begin their lazy slow-motion attack on various members of the bedstraw family, such as Cleavers, which are just throwing their new fresh growth up through the tattered crochet of

ABOVE: Bloody-nosed Beetle
The Bloody-nosed Beetle vomits a gob of scarlet, ill-tasting and smelling liquid. This is enough to put off even the most determined predator from approaching too close.

LEFT: Brimstone Butterfly
The original 'butter-coloured fly', the Brimstone male is the one we all recognize, with wings of sulphur yellow. The paler female often gets mistaken for a White. Look for the camouflaged caterpillars on the leaves of Buckthorn.

last year's debris. These beetles can be seen somewhere at any time of the month but these early warm days seem to concentrate their numbers, making them easier to see in the short growth. Whilst you have your eyes down, keep them peeled for silky drapes over the tips of the new Stinging Nettle growth as within these are the first batches of **Small Tortoiseshell** and **Peacock larvae**, getting a generation in early before the parasitic flies, which reduce their numbers later in the year, have a chance to get their own numbers up.

Like coal dust contaminating the creamy luxuriance of Hawthorn and umbellifer flowers, many insects take advantage of the instant nectar source. Dancing on the spring breezes large swarms of the black and hairy **St Mark's Fly** (named after its emergence around 25th April) hang in the air with legs dangling as if on wires. They are mostly males waiting for the later emergence of the females. Search the ground and leaves near one of these gatherings and you will find the paired up couples; males with large eye masses butting up against each other, females with them well spaced.

Easter time is the time for eggs

Eggs abound this month. It doesn't have to be edible and wrapped in silver foil to be of interest to us humans either. Try looking for the single minute eggs of newts – once you have found a weedy pond, look for leaves of various aquatic plants folded over, look in the fold and you will see a tiny blob of jelly, an individual package carefully placed there by the cloaca and back legs of the female. On a sunny day watch butterflies – their behaviour can give the game away – Small Tortoiseshells will have emerged from hibernation, fed, mated and will now be looking for the newest nettle growth to lay eggs on. Follow a female as she searches for the best place to lay her clutch, tasting with her feet. When she pauses, approach slowly. Watch as she curls her abdomen under the leaf to deposit the eggs – don't be tempted to follow her when she departs or you'll never find them. Keep your eyes firmly glued on the leaf, when you turn it over you should find several hundred tiny green eggs doing just that.

LEFT: St Mark's Fly
Breeze dancers - hanging around a Hawthorn near you are clouds of black St Mark's Fly. With their undercarriages dangling they are one of the more subtle signs of spring.

57

APRIL

Plants

It's almost as if every flower and leaf bud has been waiting under the tension of the winter for this month. All over the place this eruption of leaf and petal is obvious. **Kingcups**, **Cowslips** and umbels are among the most eyecatching. Timed to perfection many nectar-feeding insects synchronize their cycles with these plants.

The droopy flowers of the Cowslip give the plant one of its many colloquial names, 'bunch of keys' after the keys to heaven that St. Peter was said to have dropped.

Both Brimstone Butterflies and beeflies are chief pollinators of **Primroses** this month. When **Bluebells**, which for 11 months of the year are rather unobtrusive and dull, put on their well known mass flowering display, you cannot help but be taken in by the mantle of blue which hovers above many woodland floors. It is a very British thing – no cluster of Bluebells anywhere in the world gets even close to matching it. By the end of this month they are looking stupendous.

Although **Wild Garlic** isn't quite as common in such densities, it lends its own talent to the spring scene with an onslaught to the nose.

St George's Mushroom is the only large mushroom to fruit at this time of year. It is a very large fungus, with a creamy-white cap up to 15 cm in diameter. Keep an eye out for it in association with Birch trees and on old grassland. It is edible, but another genus of fungus that looks similar, *Clitocybe*, is not, so be careful with your identification.

RIGHT: St George's Mushrooms
The only large, whitish mushroom to fruit in the spring is the St George's Mushroom, which traditionally presents itself around 23rd April.

RIGHT: Pasque Flowers
It's a bit of a specialist trip unless you live close to one of the few British sites for this plant, but the Pasque Flower is one of the jewels of chalk grassland – an unsung floral tribute to this month and the next.

FAR RIGHT: Cowslips
The Cowslip is like a high-rise Primrose, a plant of open fields and downlands.

Badgers and Bluebells

Saltwells Wood, nr Dudley, (SO 932870) – A pleasant place to while away a few spring hours not far from the huge population centres of the West Midlands. A good mixed woodland whose roots swim in the blue of its beautiful Bluebells.

Herefordshire Badger Watch – In an ancient woodland setting these conditioned wild badgers can be watched from the comfort of a hide. This is as reliable as Badger watching gets. These animals are fed so they will linger and provide a bit of a Badger fest for those that attend. For more details Tel: 01989 567995, Web: www.greengate.org.uk

Cotswold Badger Watch – This is a bit of an odd one as far as Badger setts go. It's in the middle of a field! However, this allows views of the animals' behaviour without any of it being obscured by an irritating tree trunk or bramble thicket. What's more you can take friends, up to 40+ people will be tolerated! For more information phone 01453 750164. Open Apr-Sept.

Ebbor Gorge, nr Wells Somerset (ST 525488) – In spring its fresh green woodlands are cucumber cool set off against the Limestone cliffs, the white geological flesh of the Mendips. The Bluebell show here is superb along with a rich mix of all the best of the spring floral fare.

New Lanark Badger Watch – This monstrous sett has 26 entrances and is situated in a Scottish Wildlife Trust Reserve. This is a good wild experience, the only luxury being a wooden bench. However, the success rate is somewhat better than many of my attempts at Badger watching. Tel: 01555 665262, Email: fallsofclyde@cix.co.uk

New Forest Badger Watch – This has got to be a fine way of filling the gap between a day on the New Forest and a beer in the pub. This is deluxe wild Badger watching with floodlit viewing and even underground chambers. Few Badger watchers are able to get close enough they can count the hairs on a Badger's nose. You can here. For bookings Tel: 01425 403412, Web: www.badgerwatch.co.uk. Nightly from Mar-Oct, capacity 27 people.

Castle Eden Dene, nr Hartlepool (NZ428389) – Rich woodland. Rich understorey. Rich ground fauna. Probably one of the least spoilt woodlands in this part of the land, which every year between April and May fizzes with a variety of spring flowers including Bluebells, Anemones, Primroses and Ramsons. And if on your visit you see a Nuthatch then pride yourself in knowing that you've just seen one of the most northerly population of this bird!

Wormley Wood, Hertfordshire (TL317062) – A tremendous cathedral of Oak and Hornbeam rooted in an extremely varied sea of woodland flowers, including Bluebells.

Bradfield Woods, Suffolk (TL934575) – A wood against which all woods can be compared. This is a place that is managed today as it has always been since records began, at least since 1252. A place to visit at any time, but especially in spring. A quintessential British wood in which to witness the quintessential British Bluebell.

Kiln Wood, Kent (TQ888515) – A hot site for the big spring blue, this wood has a mix of old and recent coppice, creating a rich and varied mosaic of flowers. There are occasional patches of Lady's-smock, and Orange-tips can often be seen on a warm spring day weaving in and out of the Hazel stems to access the nectar or to lay eggs.

LEFT: Bluebells
No other British flower assaults the eyeballs in such a spectacular manner and this is the month that it all starts.

May

May has a wonderful full-up feel to it. It's the need to breed that is behind what May has to offer – a frantic fornication festival. Whatever direction you look, nature is in a frenzy, either fighting, feeding, breeding or flowering as every habitat makes up for lost time. From the floral explosion in every hedge comes the fuel of pollen and nectar – this, along with the warmer ambient temperature, brings on a super-abundance of insects, which in turn feeds the glut of hatchling birds, mammals and other insects.

ABOVE: Hawthorn
It isn't for nothing that the Hawthorn is known as May. Its other common name is Quickthorn. It's tough and thorny so it keeps livestock in and predators out. Birds find protection for their nests among its thorny twigs.

RIGHT: Hedgehog
Sounding disconcertingly like humans, Hedgehogs are not the quietest of lovers - hear them on your lawn or from within your herbaceous border.

Mammals

The noisy courtship of **Hedgehogs** starts this month. While out on his nighttime wanderings, when the male finds a mate he approaches her, making extraordinary snuffling and snorting sounds and moving round her in circles. This snuffling courtship can last for several hours, as the female tends to be busy searching for food and less inclined to mate.

Badgers abound
Badger watching is rewarding this month, with the cubs above ground and lots of social activity to see. But the time between when they first pop their noses out of their holes and trundle off to feed is often frustratingly short. Especially when you've been waiting quietly for hours! For badger setts to watch see the map on page 59.

Birds

The dawn chorus is in full swing, but how about making a special effort to hear some of the more adventurous entries into the songsters' chart? Although it is hard to set your eyes on the artist itself, the deep milk bottle breath of a **Bittern**, the shouting of a **Cetti's Warbler** or the drumming flight of a **Snipe** are all classics. You could seek out the immaculate notes of a male **Nightingale** and find out what Keats and Wordsworth were on about.

The crescendo of song is with us this month as many small birds sing their socks off throughout the British Isles. To a naturalist's ears this music is not a simple celebration of spring, in fact if bird-song could be translated into a human equivalent it would probably not be printable here!

Unravelling who is who from their song is a fun occupation at this time of the year, but it becomes a much more interesting experience if you keep in the back of your mind the reason that they are doing it.

Song emanates mainly from the throats of male birds as they communicate with each other and shout about their territory ownership and personal health. From these dawn and dusk performances the birds get an update on who their neighbours are. If a neighbour goes missing there will need to be a reshuffle and boundaries will need to be reaffirmed.

The peak performance is during twilight when it is simply not light enough to forage for food and, at the same time, the songster is shrouded from predators.

Females also are attracted to these acts. A male that sings is likely to have a more productive territory and needs to spend less time foraging for food. He is likely to be a healthy, fit animal and, therefore, a good breeder.

Many birds, however, do not sing. Song as a form of communication works for small woodland and garden birds because they have small territories and makes sense in habitats that are dense and prohibit visual communication. Big birds with huge territories such as birds of prey, swans and crows use visual displays as these are more efficient over long distances.

LEFT: Stone Curlew - a boggle-eyed wader!
This bizarre wading bird is very rare and limited in distribution to parts of East Anglia and Wiltshire. A visit to Cley Marshes in Norfolk can treat you to a view of this curiosity. Stay until dusk and you may hear a serenade that has to be one of the eeriest in the bird world.

RIGHT: Screaming reelers
They're back! The boomerang silhouettes of Swifts are carving up our skies on their return from African skies to the UK to breed. Joining the aerial cavalcade are those which share a similar lifestyle, but are not even slightly related: the Swallows, House and Sand Martins.

What to look out for this month

- Hedgehogs mating
- Listen for Nightingales
- Swifts arrive
- Sand Lizards go green
- Sticklebacks spawning
- Cockchafer - bug of the month
- Candelabra on conker trees
- Hawthorn in bloom

Did you know...?

If you live nowhere near a Nightingale wood and are restricted to Scotland, you have, not far away, a sound that is every bit as intoxicating as that of a Nightingale and is guaranteed to turn any southern naturalist green with envy – the eerie yodelling of **Red-Throated** and **Black-Throated Divers**, which have returned to their breeding lochs this month after wintering at sea in the south.

Breeding plumage

Non-breeding plumage (seen later in the year)

BELOW: Nightjar
Find a patch of healthy heath or regenerating conifer plantation, take a blanket, a friend and a thermos and sit out after dusk. This is a great way to enjoy the summer - to watch the display flights and hear the unusual voice of the Nightjar. Their activities are at a peak this month.

King of the singers

Now the next bit has been said a million times before in poesy. The bird I'm referring to is undoubtedly the king of all singers when it comes to the dawn chorus. In fact because of its deep roots in literature we can all name this famous bird as the Nightingale, and we all know what it is famous for. With all those quotes such as 'Singest of summer in full-throated ease' (John Keats) and 'Good nightingale! Thou speakest wondrous fair' (William Wordsworth), how can we forget?

But when it actually comes down to it, how many of us have heard one? There are some who think they have, mistakenly identifying the nocturnal bubbling of a Robin under a street lamp. That infamous Nightingale that was supposed to have sung in Berkeley Square is probably a classic example of such a misidentification.

Sadly, this bird is not as common as it used to be and the days when you could nip up to your local coppice for a listen are no longer with us. It has a range restricted mainly to the southeast and, as if this was not enough to limit most people's experience of this bird to secondhand Keats or Wordsworth, its song is best heard at night! Nightingales do sing during the day, but here they are in competition with just about every warbler, tit and thrush that fancies a go at the crown.

Some birds' eggs are hatching in May, such as **Blue Tits**. Listen out for the quiet 'tseep tseep' of baby tits calling out for their parents to bring them a tasty caterpillar for lunch.

Seabird scent

May for me is not the heady perfume of Gorse or the complicated bubbling of a **Blackcap** in a leafy lane. It's the unmistakable head-whack of guano: the odour of countless digested fish meals, normally squirted inoffensively into the sea. At this time of the year, the seabirds bring a little taste of the sea back to land with them and sugar frost our cliffs, headlands and ledges with it! To a naturalist infatuated with seabirds this bombardment of the nasal passages is as poignant a reminder of May as fresh-cut grass or wild garlic is to others.

Nightingale Hot Spots

LEFT: *Nightingale*
Nightingales are the quint-essential spring singer for some. Sadly, they are nowhere near as common as they used to be, but they are, thankfully for us romantics, still to be heard if you know where to cast your ears.

Rutland Water, Leicestershire (SK871078) – Just to prove not everything that man turns his hand to is an ecological disaster. The focus of interest here is the huge man-made reservoir with over 9 miles of shoreline, which was created in the 1970s. There is almost too much to listen for here. I must confess, though, to hearing about 7 seconds of Nightingale. Whilst I had every intention of sitting and enjoying it I was distracted by a Garganey's crackling static call. By the time I returned it had been drowned out by the songs of 11 Warblers and a Cuckoo!

Highnam Woods, Gloucestershire (SO778190) – This woodland has all the hallmarks of a pedigree ancient deciduous woodland. Glowing in the darker tangled recesses from where the Nightingales sing the flowers of the rare Wild Service Tree can be found. Turn up early for the chorus and, as light filters through, enjoy the rich ground flora, which is at its best this month.

Blean Woods, Kent (TR118611) – The aura exuded by an ancient woodland like this is no more delectable than in the cold dripping dawn of a May morning. The richness of plant habitats here is reflected by the variety of sounds present in the dawn chorus. Nightingales are particularly common, in fact this is probably one of the best spots in the country for them.

Radipole Lake and Lodmoor, Weymouth (SY676796) – Here you get two wetland habitats in one trip. These flooded grazing marshes and reed beds are superb for many spring-time favourite throatsters. It's a particularly good place to get your head around your warblers. Nightingales sing from shrub at Radipole, whilst Cetti's Warblers explode vocally along many of the walkways that edge the Reed beds. For a dawn chorus of this quality you would walk miles; here though it is all within a stone's throw of the town centre and it is wheelchair-user friendly.

Beachy Head, Sussex (TV586956) – The classic rolling chalk scape that is the South Downs ends here at its most southerly point. The open sweeping mental picture that the downs conjures up, isn't the most obvious habitat for thicket dwelling Nightingales. But they breed here in good numbers among the dense saline breeze clipped scrub tucked away in valleys and field edges.

Amphibians, Reptiles and Fish

ABOVE: Lean green sex machine
Male Sand Lizards are probably at their
most vibrant, bedecked with a bright green
flare of scales on the flanks. Accompanying
this colour change they become extremely
territorial and the presence of other lizards
can trigger all sorts of surprising reactions,
spicing up lizard watching this month.
Although restricted to southern heathland in
England, this species is well worth looking for
if you find yourself in the right habitat.

Fish, never the most watchable of British animals unless they are dangling from a rod and tackle, or the watcher is attired in diving gear, make life a little easier for the naturalist this month. For most of the year freshwater fish give themselves away only as a passing shadow or a fly-snatch ripple, remaining quietly submerged in their element. In May some species become so overtaken by the urge to breed that they lose their usual caution and find themselves breaching the interface between water and air.

Pro-creation is top of their agenda too. On a still day look out for the spawning orgies of **Common Carp**, which can be rather spectacular to watch. The violence of thrashing water and their tolerance of human viewers approaching close, contrast with their usual very quiet and timid nature. Water around reedbeds can fizz with **Roach** and **Rudd** as the males chase and pair with females.

Slightly easier to see close up are the amorous antics of that pond-dabbling classic, the **Three-spined Stickleback**. The male's belly becomes flushed crimson, his eyes blue and he becomes super-aggressive not just to any male invading his 'territory' but to just about anything you care to place in the water that is red.

Lift stones from the stream bed and it is not uncommon to come across a mass of pink eggs attached to the stone itself or the river bed.

BELOW: When Sticklebacks attack ...
The Three-spined Stickleback is a fish often
caught in children's pond nets in the spring.
They're small, but aggressive. Don't forget
to put them back afterwards.

LEFT: Rudd
When the water appears to boil in slow-moving and still water this month it is often this common fish, the Rudd, that is causing the commotion. Look out for the red fins, but be careful, it is very similar to another species the Roach.

Nearby should be an attentive male **Bullhead**, or Miller's Thumb, a small, secretive but common river fish, which again seems only to be known among the angling fraternity, who occasionally snag one by mistake, and kids who dabble with nets in the fast and stony sections of stream or river.

Of all the small fish spawning now probably the most mysterious and impressive in spectacle are the writhing spaghetti masses of **Brook Lampreys**. These fish look a bit like mutant eels, with a long sinuous body, a sucker disc of a mouth and seven gill holes making it look like it has been spiked in the head with a fork. The spawning is really the culmination of a five-year lifecycle; the juvenile fish, known as prides, hoover the stream bed for detritus, then, as they turn into adults, their gut dissolves and their last act before they fade out is to spawn.

The other schoolboy's 'tiddler', the **Minnow**, goes through a metamorphosis like the Three-spined stickleback; the males develop white tubercles on the head and gills and their bellies go pink, while the breeding females swell with eggs soon to be strewn on pebbles and stones on the stream bed. They seek the spring-warmed shallows. Look for them congregating in large shoals for spawning. These fish, often completely overlooked despite their importance as food for other fishes such as trout and salmon, also fuel many a nestling Kingfisher's development about now.

BELOW: Minnow
Often overlooked, but actually one of my favourite little fish, the Minnow's usual role in life is as dinner for something bigger! In their own right, especially when the males are in their breeding flush, they are stunners.

Practical bit

A fish in the jar
You can get good views of Minnows if the water is clear and calm, but a neat little childhood trick known to me as a 'Jilly jar' can be useful. Simply tie a piece of string around the neck of a jar; bait the jar with a crust of bread and submerge it in water neck away from current; let it sink – then when lots of curious fish have entered pull it rapidly to the surface. Hopefully, you will have caught a few. It is now you can appreciate what a pretty little fish it is with a variety of different colours from bronze to white with dark banding.

Invertebrates

Look out for the early workings of **Leaf Cutting Bees**, which will be starting to dismantle rose foliage. Hovering above either a Primrose or the entrance hole of a solitary bee you may notice the fluffy form of the **Bee Fly** often first detected by its high-pitched humming.

ABOVE: Rose Chafers True jewels, these beetles sporting burnished metallic armour are attracted to artificial lights. Look out for them on windows, porches and, in good years, littering the pavement beneath street lights.

RIGHT: Swallowtail Arguably our largest butterfly, it certainly must get the prize for being the most exquisite. Large, powerful and with two tails on its hind wings, which give the Swallowtail its name.

Bug of the month

This title just has to go the **Common Cockchafer** or Maybug! A fresh one is covered in a downy golden pelage which clings like threadbare velvet to the chestnut-brown wing cases and the rest of the body, the underside of its belly and its big eyes are a glossy black. If ever an insect can be ascribed a personality then this is it – a real Biggles of an insect, even if it is a little clumsy on take off and landing. Of the three species most often encountered (the metallic green **Rose Chafer** or the tiny tan and green **Garden Chafer** are the others) the commonest is the aptly named Maybug (due to its first appearance dates) or Cockchafer (because it is the biggest presumably).

If you have never seen one of these chiefly nocturnal insects you can do a lot worse than hang around street lamps, as they are attracted to light. The beetles themselves have spent the last two to three years as C-shaped larvae (also known as rook worms owing to the apparent fondness of the bird for them) and would have pupated and then emerged as adults back in the autumn. They have been waiting for the relative warmth of May and June to take to the air and find a mate.

If you find a Maybug, sit it on the top of your finger and, as it gets active, watch how it pumps its abdomen in and out, circulating its body fluids and generating heat by muscular activity. When it is up to the right temperature (about 38 °C) there is a pregnant pause as it lifts its wing cases, unfolds its transparent flight wings and … in about half the instances then proceeds to plummet to the ground! After several attempts this unlikely looking aeronaut finally buzzes off.

Swallowtail

It has to be the sexiest insect in the British Isles. It has good looks, confidence and the added allure of being exclusive to the Norfolk Broads that associate with the Rivers Ant, Thurne and Bure. To watch a **Swallowtail** glide genteely between the reeds and rushes or skim in tandem with its reflection across the surface of one of these ancient waterways has got to be close to most lepidopterists' ultimate fantasy. It combines size (it's arguably the largest British butterfly) with a stunning coloration, lackadaisical flight and rarity.

If you have luck on your side and get a warm still day at the end of this month and find yourself in this region of England, you get a lot of insect for your effort. I have actually found it rather easy to see, certainly compared to its rival for the title of largest British butterfly, the **Purple Emperor**. It's worth bearing in mind as you watch its seemingly care-free flutterings, that despite depositing a large number of eggs on the leaves of Milk Parsley this is where the butterfly's idyllic life ends as over 65% of all the caterpillars that hatch are consumed by predators, mainly spiders.

Shield bug lifecycle

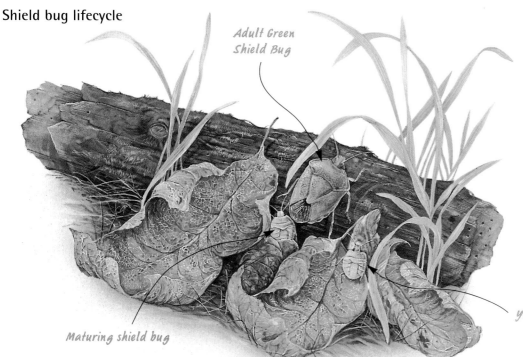

Adult Green Shield Bug

Young shield bug

Maturing shield bug

LEFT: Shield bugs
Now that the air temperature is rising, things are in full swing in the six-legged world. Look out for Green Shield Bugs.

Two quintessential spring butterflies, the **Pearl-bordered** and **Small Pearl-bordered Fritillary**, lay their eggs near or on the Dog Violet's heart-shaped leaves. At the same time, quietly putting away violet leaves are the caterpillars of two of the same family of butterflies – the **High Brown** and **Dark Green Fritillary**.

The first of the summer's hawk moths are on the wing. Both **Eyed** and **Poplar Hawk Moths** are dead ringers for dead leaves, until disturbed and then they'll flash eyespots to make you jump.

Beetles of doom

With nature's floral tribute to spring comes added interest for those with entomological tendencies! Put simply, flowers attract insects. As anyone who cycles along a Cow Parsley-lined hedgerow with their mouths open will know, this plant attracts hordes of pollen and nectar junkies. Stop and examine the landing platforms of dense flowers and you will witness an orgy of mastication and ingestion as the glut of flies and beetles exploit this ephemeral resource.

Look closer and you will notice a number of small beetles, boldly bedecked with chimney reds, oranges and navy greys. These are the **Soldier** and **Sailor Beetles**. There are four common species: *Cantharis rustica* has blue-grey wing cases and a spot on the pronotum (plate-like part of the insect's body that sits over the thorax), *C. fusca*, which is similar but with black legs, *C. livida*, which has brown wing cases, and

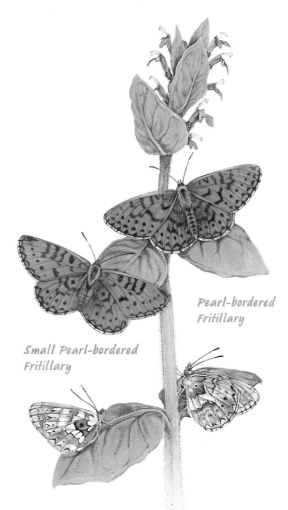

Pearl-bordered Fritillary

Small Pearl-bordered Fritillary

LEFT: Pearly queens
These two butterflies, the Pearl-bordered and Small Pearl-Bordered Fritillaries, are found on the wing only in spring. As they are fairly similar in appearance, it takes a bit of practice to separate them.

what, as a child, I used to refer to as a blood sucker *Rhagonycha fulva*, which is redder than the rest with orange wing cases with a black band towards their ends.

Watch them for a while and you will notice they display not a penchant for pollen but a deathly disposition as they chomp any insect that stays long enough.

Umbellifers and Hawthorn flowers are also the chomping grounds of another beetle – the crimson **Cardinal Beetle** – whose taste is slightly more conventional, attracted to these flowers for their nectar.

RIGHT: Spiny Spider Crab
This would be quite a find for the crab enthusiast! These common crabs live all around our shores, but we usually see them only after a storm, when they tend to get smashed up. This shell, though, is perfect and well worth collecting.

Shell out

Go for a walk on the beach this month and you may well notice what appear to be lots of complete dead crabs. Most of these are not whole animals but on closer inspection are hollow shells. **Crabs** need to moult in order to grow, like other animals with their skeleton on the outside. Female crabs need to moult before they mate. This amounts to a lot of crab shells lying about.

Even though you may not get to see the live ones, collecting, drying and gluing these to card allows you to get to know many of our British species. You can add to your collection as you travel around and visit beaches.

Ant attack

Wood Ants nest in those large heaps of twigs, leaves and pine needles you often find in open woods. These little insects will do anything to defend their nest.

Get this trick wrong and you'll know all about one of their weapons – their jaws. They bite and it hurts! But they have a secret weapon – the acid pore! Place your hand close to the surface of the nest and you will notice some of the ants do weird things, sticking their bottom between their legs. Now smell your hand – it should niff of vinegar. This is formic acid, which the ant squirts at attackers from a tiny hole near their bottom – an acid water pistol!

If one ant starts, soon all its mates join in. In fact, carefully place a short lighted candle amongst the insects and they will soon put it out with their spray.

LEFT: What a squirt!
Rock pools are really happening now! Look out for colonial seasquirts. These unusual animals form gelatinous blankets over rocks around the low tide mark. Keep your eyes open for a strange snail, the European Cowrie, which can sometimes be found grazing on them.

Plants

At a distance mauve smoke seems to drift and circulate around unfurling fronds and grass spikes and the dead crisp of last year's **Bracken** and grass stems. Look a little closer and you will notice the pigment radiates from a familiar face – looking like the Pansy of hanging basket and ornamental border fame, **violets** belong to the same *Viola* family. There are 13 species in the UK, all of which can be found in flower this month, these include the colourful source of the all the designer varieties, the **Wild Pansy**.

But despite the variety of species and all the subtleties of identification, none seem to represent the month more than the **Dog Violet**. It can occur as a quiet individual or a huge explosion of colour that turns a rough grassland, Bracken slope or woodland floor into a purple blur.

BELOW: Dog Violet
A member of the same family as pansies, the Dog Violet is our commonest violet, but it is not scented.

ABOVE: May mauve
May is a month of mauve and blue. Roadsides and meadows all over can abound with the purple spikes of the Early Purple Orchid (shown here). Not quite so in-your-face are the secretive and rather reserved displays put on by Solomon's Seal in ancient chalk woodlands. A mass spectacle to be seen in a few localities around the country, are Snake's-head Fritillary flowers nodding in the breeze in a few select meadows.

Oak flower

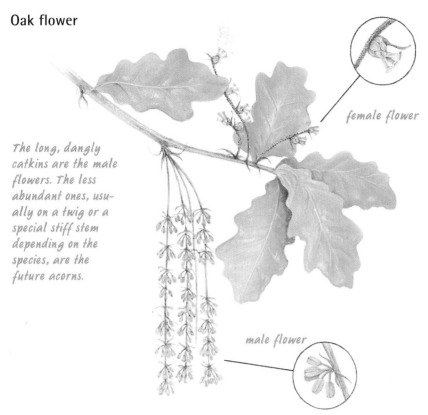

female flower

The long, dangly catkins are the male flowers. The less abundant ones, usually on a twig or a special stiff stem depending on the species, are the future acorns.

male flower

Horse Chestnut flower

These very ornate 'candles' of flowers are one reason for the tree's success. Horsechestnuts were once planted for their ornamental value during their short, but unmissable, flowering period.

If the flowers are pink and white they are White Horse Chestnuts. If they are dark pink they are Red Horse Chestnuts (an ornamental hybrid).

Tall flowers, high bloom

Like trails of fizzy shaving foam cast over our landscape, Britain's most popular hedging tree, the **Hawthorn**, is in full froth right now, giving sense to its other popular name of 'May'.

With towering white candelabra of flowers, **Horse Chestnut** also provides a bounty of pollen and nectar to the myriad insects on the wing. In complete contrast, a tree which has associations with a very different season, **Holly**, is in flower now too. Check out your local tree for clusters of tiny four-pointed star-like flowers. Look even closer and you can see that flowers on each tree belong to one sex or the other. Male flowers hold their anthers high and while females have dysfunctional male parts they also possess a three-lobed stigma smack bang in the middle. The fact that you have male and female trees explains why some trees never produce berries at the desired time of the year, when we all go foraging for Christmas decorations.

All these trees are pollinated by insects and that is why we should have no trouble noticing their displays, even from a distance. They are designed to be a billboard, advertising their wares to the insect world.

Trees that rely on a breeze to courier their goods from male to female flower, tend to be a bit more of a challenge. **Oak** produces male catkins, which, while not as obvious and yellow as trees such as Hazel, perform the same function: to launch pollen into the air and away from the tree. You most often notice these when they have shrivelled and dropped from a height, especially if you park your car under one at night – they have a habit of getting stuck under your windscreen wipers!

Female flowers are less showy and are contained on a short stem or grow from the twig depending on the species of oak. A good urban tree that goes through the same actions, but rarely to any effect, is the **London Plane**, that handsome tree with flaky bark that lines many pavements in our cities. Planes produce male and female flowers in those well known pomander-shaped affairs. They are, however,

TOP LEFT: Oak flowers
You may not think of Oak trees flowering, but they do. How else would you get acorns in the autumn? Look for the catkins.

LEFT: Horse Chestnut flowers
Conkers to come. These pink and white lovelies are hard to miss in May and June.

ABOVE LEFT: Common Gorse
Little can compare to the heady, heavy sweetness of air laden with the aroma of fresh gorse, especially on a still, breezeless day. The insects seem to agree too!

ABOVE RIGHT: Hedge fluff
The frothy flowers of Cow Parsley come alive along with other flowers and are accompanied by an ensemble of insect pollinators.

BELOW: Big head
One of the larger umbellifers, Giant Hogweed is a common hedgerow flower, standing grand among its smaller relatives.

unable to set viable seed owing to their mixed parentage. London Plane is not only a non-native but also a hybrid between a North American species *Platanus occidentalis* and *P. orientalis* and is sterile.

Parasols of petals

It is a confusing family of plants, the Umbellifers. The name refers to the platform of flowers supported by green spokes resembling an umbrella – so that is easy to remember. But it is such a diverse family that separating the many different species can be a bit of a botanical nightmare. What is clear is that they are hugely evident on just about any bit of roadside verge, hedgerow and wasteland at present. From the towering **Giant Hogweed** to the intricate coy inflorescence of the woodland **Pignut**, they are all flowering at the moment.

Probably the commonest and the most dominant of the family is **Cow Parsley**, a rather dull name for a plant that adds its spectacle to late spring, frothing out from the bases of many a country hedgerow. Its flowers are the ideal landing platform for many insects and as a direct result of this, panting country cyclists would be wise to keep their mouths shut on passing a clump!

June

June is definitely the month to head to your favourite cliff-top birdwatching site for the first time in the season. The smell that wafts over the headland takes you back to the last time you were here. You simply cannot fail to have a good time at a seabird colony, there is so much going on. It doesn't matter where you go and what you watch, but head for a headland, any area where there are cliffs, and you will almost certainly have seabirds of some kind nesting there. Fulmars, gulls, perhaps Gannets and, of course, the auks – Guillemots, Razorbills and Puffins – all provide a fantastic backdrop and murmuring soundtrack to the whole experience.

ABOVE: Cliff watch There is no place I would rather be on a sunny summer's day. It's all here! The world of birds is particularly rewarding – just about everything is nesting and doing something interesting.

Mammals

True water babies are being born this month betwixt the tides. Seals in general have the aquatic lifestyle very well arranged but they must return to shore to give birth. Of our two British species it is the **Grey Seal** that fulfils most people's image of a baby seal – a white fluffy lump of blubber with limpid eyes. But an avid peruser of pinnipeds will be disappointed if this is what they expect to see; Grey Seals only give birth during the winter months.

It is the **Common Seal** that is at the peak of its birthing right now and, surprisingly, of our two species it is the rarer. Common Seals number around 25,000 and breeding populations tend to be concentrated in certain areas. Hot spots include: The Wash in England; in Scotland they are found in biggest numbers in the west

with some on the east coasts; and in Ireland they are scattered thinly around the north and down the west coast.

The reason you won't see fluffy babies is that the Common Seal pup is born a sleek, lean swimming machine, complete with the fine, waterproof hair of the adult. The fluffy coat of hair (known as lanugo) is moulted off in the womb. This enables the females to give birth on sand banks and rocks exposed at low tides.

Fox watching
Most mammals have just experienced the annual baby boom. Now while a few of us will be lucky enough to see young **bats**, **Hedgehogs**, **Common Seals** or **deer fawns**, most of us will not. Try foxes.

Nearly all of us will have **Red Foxes** living near us, whether in town or country. All you really have to do is find where the breeding

LEFT: Common Seals
Look out for Common Seals hauling out around our coasts. The young are born this month. They are not the fluffy little baby seals of postcard fame, but full-on water babies.

BELOW: Red Foxes frolic
Signs of fox cubs become more and more evident as these creatures get bigger and even more adventurous and frisky.

earth is. This is not as hard as it sounds this month, simply think fox. Small copses, a stand of trees in a park or that tucked away corner of wasteland are all good places to start. You are generally looking for a single hole (although they will cohabit with **Badgers**, which can be a little confusing) that is bigger than that of a **Rabbit** but not quite the size of a Badger's, without a huge mound of excavated soil.

The signs of fox cubs are very recognizable. Leave young humans in a room on their own and they will destroy the furnishing and drag all their toys out and leave them discarded as soon as they are bored with playing with them. Fox cubs are the same.

Vegetation around the earth will be flattened. With all the rough and tumble, there will also be many playthings lying around – bones, feathers, whole wings, rabbit's feet and, in urban situations, objects such as rags, flowerpots and balls – stolen from gardens by the adults. On top of this the whole area will smell strongly of fox.

Once discovered, it's a simple matter to stake out the location. If it is a quiet spot, quietly stalking the area on a warm June afternoon may pay dividends. Fox cubs, impatient for the cover of dusk often emerge above ground early to bask and play. Failing this, early morning and dusk visits will almost certainly be good. In these circumstances, you often hear the cubs before seeing them as another thing they have in common with their human equivalent is that they can be excessively noisy!

What to look out for this month

- Frolicking fox cubs
- Terns inland
- Cliff-nesting seabirds
- Froglets and toadlets
- Speckled Wood butterflies
- Elderflowers
- Grasses

Birds

It is not just the mammals that are going forth and multiplying. Birds are going great guns and nowhere is this more noticeable than in a seabird colony, where the clamour, stench and incredible variety soon causes one of the best British sensory overloads you can experience. But if heading for the rocky sticking out bits of the UK seems like a bit of an effort, it is not always necessary, for plenty of ornithological action can be experienced at your local gravel pit or reservoir – in the form of **terns**.

Juvenile - white forehead and dark wing markings

Adult - scarlet bill with a black tip

Terns combine grace and elegance (the tern family were often called sea swallows) with a loud extrovert boldness that is found in few birds in the world, as anyone who has visited the Farne Isles without serious head protection will no doubt confirm.

There are five British species to see. However, create a large body of water, with nice gravelly beaches, and us humans tend to appear in our droves on Sundays, make a lot of noise, throw sticks for dogs and race about on jet skis, and generally make life hell for a bird that likes a nice quiet scrape nest in the gravel.

Then an enterprising naturalist came up with a good idea and for a number of years, floating rafts of gravel at numerous inland sites, have created a sanctuary for terns. A very pleasurable time can be had watching these birds chattering to themselves, hovering with deep wingbeats, bills to water before diving through their own reflection to take Minnows from the shallows. What is extra special is that you can watch the rest of the trip as the adults return to their chicks bobbing away on their raft.

ABOVE: Common Terns
It is the Common Tern that is most frequently encountered inland. The young bird has a white forehead and scaly back.

LEFT: Taking terns
Harbours and warm coastal waters bring plankton and small fish like Sand Eels into the shallows - fodder for the plunge-fishing terns. Watch them and it is easy to see why they got the name 'Swallows of the Sea'.

Terrific Terns

Farne Islands, Northumberland (NU230370) – If you fancy an introduction to seabirds in general or want to see all the British terns except one, you cannot pick a better place. A visit to the Farnes has got to be one of the best natural history experiences in the world.

Lough Neagh, Oxford Island, Northern Ireland – Although there are no easily viewable breeding colonies in Northern Island, they are present mainly on the little disturbed islands. A trip to the visitors' centre on the edge of this large freshwater lake will guarantee good sightings of these dapper dippers.

Rye House Marsh, Lee Valley (TL386100) – You want junction 25 off the M25. This is one of the best spots for a good day's birding within easy reach of London. There are also lots (somewhere in the region of 100) of other water bodies in this area, all worth checking out. Rye House Marsh (RSPB reserve) is one of the best. Excellent for views of breeding Kingfishers, Sand Martins and Common Terns on rafts all viewable from a hide.

Abberton Reservoir, Essex (TL961185) – This particular spot is famous for its tree-nesting Cormorants, but also from the hides near the visitor centre Common Terns can be seen at the nest, and their constant bickering ensures you cannot miss them.

Brownsea Island, Poole Harbour, Dorset (SZ032877) – The lagoon on the island has its water level managed so that the artificially-created islands are not flooded and this has resulted in colonies of both Common (internationally important numbers) and Sandwich Terns (most westerly site in the south).

Stodmarsh, nr Canterbury (TR222607) – This is a good place to see Common Terns which have benefited from the rafts built for them in the main lagoon. The whole place is crawling with warblers, particularly the reedbed and alder carr. Good views of Hobbies are often had from the flood defence wall – best on still summer evenings, when they hawk insects and martins over the water.

Rye Harbour, Kent (TQ942187) – If you ever had a user-friendly name for a location then this is it. The Ternery Pool is where the terns are! Both Common and Little Terns breed here on the shingle with occasional attempts by a few Sandwich Terns. All the action you could want to see is overlooked by a couple of hides.

Juvenile - distinguished by scaly feathers and dark bill

Adult - with clear head markings and yellow bill

LEFT: Little Terns
Once one of Britain's rarest seabirds the Little Tern has benefited from private nesting beaches and concentrated conservation efforts.

RIGHT:
Piles of Puffins and oodles of auks
Seabirds come home to nest and can be seen in their plenty on steep and inaccessible cliffs and islands that afford protection from any predators.

BELOW:
Razorbills
These Razorbills are in fine fettle and are just an example of the delights that can be seen this month.

Love on the rocks

It is time for the big bird spectacular on many sheer, rocky and sticky out bits of the British coastline as they become the crowded tenements for many seabirds that come ashore for a short appearance about now. Even though the first beaked whispers could be heard in early spring, now the cacophony reaches its climax.

Those 'northern hemisphere penguins', the **Common Guillemots** and **Razorbills**, will be in the full swing of breeding. If you get your timing right, it is possible to see all stages: adults, their single exaggerated pear-shaped eggs, the gauche fuzzy nestlings and the more streamlined pre-fledging chicks. Within about 15–20 days of hatching, before they are able to fly, these 'jumplings' evacuate the nest ledges and head out to sea with the adults. A visit after this has occurred will leave you with a rather paler impression of the cliffs. Early nesters such as **Cormorants** and **Shags** will still be bringing up their gawky chicks. **Puffins** stick to their private roof top apartments, nesting in burrows among their own landscaped gardens of flowering white Sea Campion and pink Thrift. Deep in the crevices and boulders at the base of some of the more northern colonies nest those slick characters, the **Black Guillemots**, in their black, white and red finery.

If you find yourself getting telescope eyes you can always lie back and take in the wonderful sea view to the soundtrack of **Yellowhammers**, **Skylarks** and **Whitethroats** singing – that is if you can resist scanning the twinkling mass of water before you for dolphins or Basking Sharks!

Be warned! Cliffwatching is one of the most addictive summer pastimes. Just as you pack up to leave, something will happen. A **Peregrine** will do a fly past or that **Lesser Black-backed Gull** will finally catch a nest off guard. So here's

Don't approach a nesting adult - it may spit a foul-smelling oil at you.

LEFT: Fulmar - Albatross of the Atlantic
Once not found on the mainland coasts at all, the Fulmar is now very widespread. Listen out for its noisy nesting behaviour and look for its stiff-winged flight.

some advice (and I speak from experience): it is prudent to have a few well-rehearsed excuses up your sleeve to explain why you were late home for dinner.

If you are unable to visit one of the great bird colonies, it is worth looking out for the 'albatross of the Atlantic', the **Fulmar**. It looks a little like a gull, however, look for the flying 'cross' made by its stiff level wings and its grey back and tail. No gull has these features and if you get close enough to see its black ink-smudged eye and tubular nostrils, there is no mistaking it.

The nice story about the Fulmar is that back in 1878 the only place you could have seen one was on St. Kilda. Now its population has expanded and it has colonized many of the suitable cliffs around our coast, preferring those with soil ledges rather than bare rock.

Fledglings

This is the time of year to keep an eye out for fledgling birds, such as baby **Blue Tits**. They will be leaving the nest and setting out on their own to find food.

BELOW: Thrift
If you fancy a break from birding, a bit of botanizing on the cliffs will reveal many specialist plants, such as this Thrift, most of which are known as Halophytes, in other words, salt-tolerant.

Did you know . . . ?

Puffins are often called Noddies after their nodding habit! In the Farne Islands, the seventh century home of St. Cuthbert, they are called Cubby Noddies. They are also referred to as London Tourists because the adults seem to spend much of their time sitting around looking vague!

Rocks and Rock Pools

ABOVE: Gannets
It is a luxury, but there are places where you can get to witness a vista such as this. Do your homework and you will be rewarded.

Bass Rock, Lothian (NT602873) – This 90-m tower of volcanic rock appears 'iced' with birds and white guano during the breeding season. This has to be a highlight of the British coastline for its 18,000 breeding Gannets among others. Various boat trips are run from nearby North Berwick.

The Farne Islands, Northumbrian coast. (NU230370) – An ornithologist's fantasy island, Over 19 species of seabirds breed here including Puffins, Guillemots and terns in their thousands.

Bempton Cliffs, Humberside (TA196740) – One for all lime-stone lovers, these are not only the highest chalk cliffs in Britain at 120 m, but they are home to the only mainland gannetry. As well as all the other cliff-nesting birds, a rich flora is present and this in turn attracts good numbers of butterflies.

Rathlin Island Cliffs, Co. Antrim (101512) – This is the largest seabird colony in Northern Ireland. As well as all the classic cliff-ledge birds, there can be good views of Black Guillemots and Eiders. Booking your visit with the RSPB warden is essential.

Skomer Island, Welsh Coast (SM725091) – This island is the biggest in a cluster and they have it all – birds, cliff plants and rich shore life. The blankets of Thrift and Sea Campion have got to be seen. A boat leaves regularly from Martin's Haven.

Wembury Marine Conservation Area, Devon (SX518848) – Good spot for Cornish Suckers and a host of other rock pool fish including Lumpsuckers, Pipe Fish, Blennies and young eels. There is a visitor centre which complements any visit with a range of displays relevant to the avid rock pooler.

Goodrington Sands, Torbay (SX893583) – Right in the middle of the English riviera is some of the best rock pooling to be had. To help you out there is a seashore centre, a rock pool exhibit, organized events and smiley staff to answer any questions.

Kimeridge Bay, Purbeck Dorset (SY909788) – This, the oldest voluntary marine reserve in the country, has a long period of low tide and is a top spot for rock pooling. The visitors' centre is a good spot for beginners with displays, aquaria and a touch pool to get you started.

Amphibians, Reptiles and Fish

While you are near the coast you may well be tempted down to the rock pools. Take with you a sacrificial meat sandwich to get glimpses of one of our commonest rock pool fish the **Common Blenny** or Shanny. Place a portion of ham or any other meat in your chosen rock pool with plenty of crevices and weed and you should not have to wait long before you are visited by this red-eyed rock pool specialist. This remarkable fish has a hidden talent: if it finds itself a fish out of water, stranded by the tide, it can use its large powerful pectoral fins to 'walk' to the nearest water. It can survive some time out of the drink aided by a thick mucous-covered skin, which slows down the evaporation of water from its surface.

Britain's most bizarre fish can also be revealed by searching those rock pools on our western and northern shores. The **Cornish Sucker** or Clingfish makes up for what it lacks in size with its excessive weirdness. It is small, up to 7 cm long, and rather extravagantly coloured with reds, greens and two large blue eye spots on its head, with a mouth that is prolonged into a snout. During this month Cornish Suckers are spawning and a good way to look for the batches of their yellow skittle-like eggs, stuck to the underside of rocks and guarded by one or both of the adults, is to use a hand mirror (a magnifying one is useful) on the end of a stick or wire. This will enable you to look under

rocks and into crevices without constantly lifting up rocks and weeds, or grazing your knees on barnacle-encrusted rocks.

Reservoir frogs

Where there were masses of Common Frog and Toad spawn earlier in the year, now there should be masses of **froglets** and **toadlets**. Head down to fresh water where spawn was present and scan the surrounding vegetation with a torch.

ABOVE: Common Frog
Newly emerged froglets are often smaller than a thumbnail. They ping from damp pondside vegetation. Very few will survive to adulthood.

LEFT: Cornish Sucker
This fish gets its name from a large and complicated sucker disc created by the pelvic fins on its belly. This is so effective that, if a specimen is caught in a net and you pick it up, no matter how vigorously you shake your hand, if the fish doesn't want to let go it won't budge. This and its low, rock-hugging profile are an adaptation to its often rather turbulent surf-swept lifestyle.

ABOVE: Common or Smooth Newt
Finding a newt out of water means it is probably on the hunt for food or on its way to a dark, damp spot in your rockery.

You will discover a front line of these little hopping troops making their mass exodus from water to hide up in the long damp grass. It is a little harder to identify which species you are observing as both appear black and shiny; the giveaway is that toadlets have tiny golden spots on a jet black background.

They have one task – to eat and remain uneaten. The long, rank vegetation provides both food, in the form of tiny invertebrates, and shelter from predators such as birds in the day and Hedgehogs, rats, foxes and Badgers by night, all of which are ravenous, with young to support at this time of year.

If you wish to witness the whole saga, sit downwind using a powerful torch with a red filter and watch. Every animal knows of this bounty and will call in on their nightly rounds to Hoover up some unsuspecting amphibians. Seeing a Badger sweeping its snout from side to side, as if it were a metal detector, it is easy to understand why so few of these froglets will survive to hibernate.

While you are in the vicinity of the pond keep an eye open for other adult amphibians including **Common (Smooth)**, **Palmate** and **Great Crested Newts** (the last is fully protected by law and should not be disturbed). It is often a surprise when you first find them out of water, but after breeding in the pond in the spring they emerge on to the land to feed, spending much of the daylight hours for the rest of the year lurking under logs, stones and deep among the vegetation.

LEFT AND ABOVE: Catching newts
Disturb the weed around the warm edges of your favourite newt pond and you will uncover the little tadpoles of the newts you watched courting at the beginning of spring. It looks just like a newt but smaller. Those fluffy things sticking out of the side of its head are gills.

Invertebrates

With warmer days being conducive to a beach visit, look out for **Common Shore Crabs** in berry. These are females which are carrying their batches of salmon pink or brown eggs around with them under their tail flaps. Do not mistake these for the yellow Parasitic Barnacle *Sacculina carcini* which looks similar. Infected crabs cannot moult and often stand out as their carapaces are encrusted with tubeworms.

LEFT: Very in berry
This is a Common Shore Crab, which has been turned upside down to reveal the mass of eggs held to her belly.

Jaws

The first time I saw this animal it was a crazed, flattened, tessellated jigsaw of bits – all stuck together with the goo of its own haemolymph. The only salvageable fragment was the left mandible. It was enough to intrigue; terrible looking, serrated with the colour of polished cherry wood. I would return.

That warm June night, in south London I did just that, and waited beneath the lamp post that was nearest the pavement on which I had found the smashed male **Stag Beetle**.

At half past midnight my reward clanged into the light and hit the ground spinning. It was a female about 3.5 cm long, lacking the overblown mandible 'antlers' used by the male in fighting, just like deer stags. Hers were stout and sharp as tin openers – functional, but, like the males, not used for actually eating (they imbibe sap oozing from trees with a short yellow tongue). They are used for excavating soil and rotten Oak wood, in which to lay her eggs. Here the white padded larvae will chomp away for up to five years before emerging as the largest beetle in Europe.

Stag Beetle lifecycle

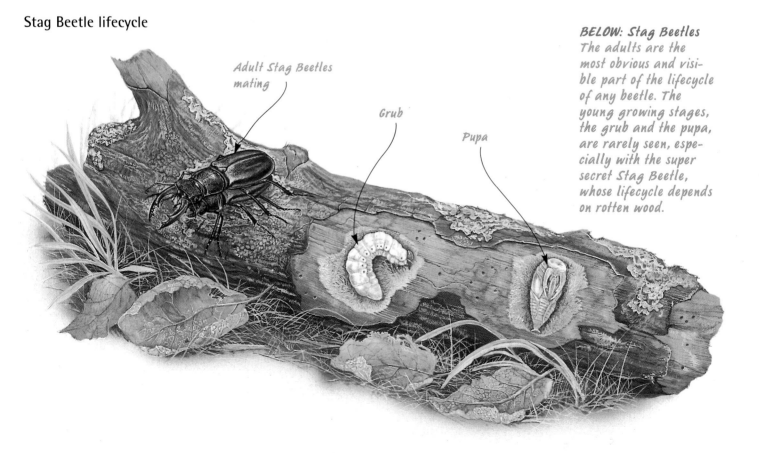

Adult Stag Beetles mating

Grub

Pupa

BELOW: Stag Beetles
The adults are the most obvious and visible part of the lifecycle of any beetle. The young growing stages, the grub and the pupa, are rarely seen, especially with the super secret Stag Beetle, whose lifecycle depends on rotten wood.

RIGHT: Lesser Stag Beetle
A smaller beast than the Stag Beetle, the Lesser Stag Beetle is still a big and spectacular insect. Neither sex develops those big over-the-top mandibles or 'antlers' of the male Stag Beetle.

It wasn't until two Junes later at the same lamp-post that I was rewarded by the sight of the awesome male, a specimen that measured 6 cm from tip to tip. They are an increasingly rare animal now in this country. I saw that male eight years ago. I haven't seen one since.

Scorpion Flies

I know I'm pushing it. Trying to identify flies is difficult enough let alone getting to know them. But the **Scorpion Fly** is different – for a start it isn't a true fly but a member of the order *Mecoptera*, more closely related to the Lacewings than to flies, and of all the winged insects they are actually pretty easy to identify.

Search damp hedgerows or even beat them with a stick over a white sheet and you are pretty certain to find these flies. They are weak fliers and fairly sluggish movers, preferring to walk around in their search for food or mates. Their name refers to the male's sexual organs, which are bulbous, carried at the tip of his body and often held curled up over his wings. This,

Practical bit

RIGHT: Pond dipping
Stare into a net full of the pond's secrets. Always, and I mean at any time of the year, this is a rewarding place to find wildlife. Dragonfly and damselfly nymphs are the top bugs to hunt for this month.

Dragons – fly!

We all know about rearing caterpillars and watching the butterfly emerge from the chrysalis – but have you ever thought about doing the same with **dragonfly nymphs**?

Get down to your local pond and fish around with a net. Look out for the large, creepy nymphs of these insects (use a book to help you if you are not sure what they look like). Select the biggest specimens to take home.

Set up individual tubs for each nymph outside – ice cream containers are best. The more you have the better your chance of witnessing the miracle. Place plenty of weed and other smaller pond creatures (small worms are a good substitute) in each tub to provide food and shelter and then, most importantly, a nice branched twig for the nymph to climb up.

Then wait, check the nymphs regularly every evening and early morning and you could see the moment when the nymph crawls out of the water, splits its skin and emerges as an adult dragonfly. Once it's dried its wings it will fly away. Don't worry if you miss it. The dragonfly will have left you a souvenir to keep – its old skin!

RIGHT: Scorpion Fly

RIGHT: Scorpion Fly
Looking like it can ruin your day, the dozy,
slow-flying Scorpion Fly is actually harmless.
This seedy denizen of the hedge and garden
is a scavenger of the
dead, wounded and
helpless. The 'sting'
isn't one at all but the
complicated reproductive
organs of the male. The female
doesn't have these.

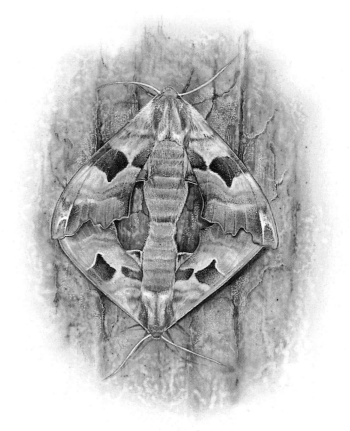

combined with his rather bold yellow and black markings and long beak-like mouthparts, will often leave the human discoverer fleeing with arms flailing. But despite their looks they are harmless, feeding on a catholic diet of dead and decomposing plant and animal material – even stealing from the webs of spiders!

The males are true 'gentlemen', offering gifts of saliva or guarding a food item from other males while emitting a pheromone to attract a female. Once she is satiated, mating commences.

Beautiful butterflies

The first wave of true summer butterflies will fly this month. Along with the first generation of **Common Blue** and **Speckled Wood**, look out for a moth-like butterfly, the secretive **Dingy Skipper**, and the extravagant butterfly-like moth, the **Cinnabar**. Truly nocturnal, it is easily disturbed and often seen displaying its scarlet flashed dark green wings during daylight.

RIGHT: Lime Hawk Moths mating
Lime Hawk Moths are probably the most fre-
quently seen of the hawk moths since their
caterpillars feed on Lime and various species
of Poplar and Willow. This brings them into
our gardens and city centres. Look for fresh
adults getting jiggy with it on tree trunks!

LEFT: Woodland warriors
The perky Speckled Wood butterflies are on
their second generation of the year this
month and a walk through dappled wood-
land will often put up the males, which,
after a brief aerial sortie, will return to
their sun spot. Sit and watch and you will
see that the bold butterfly will give chase to
any trespassing insect. An encounter with a
male or female will result in a vertical
tumbling flight.

Plants

Of all the plants I could mention this month, many of which are at their flowering peak and easy to see, here are a few weirdos that are not attractive in the conventional sense.

Rape, murder and pillage
Anywhere you can see **Broom**. Look for the tall, anaemic, fleshy coloured, flower spikes of the parasitic **Greater Broomrape** – this true life 'triffid' grows on the roots of its host, perennially sending its flowers above the surface. Similar is the smaller yellow tinged with red flower of **Common Broomrape**. It is commoner but it doesn't rape Broom; instead members of the pea family are its preferred host.

If you find yourself on moor, wet heath or bog, investigate any areas of exposed wet soil. Here you may find the tiny lit-

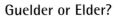

ABOVE: Death by dodder
Yes, plants can be nasty too. The Common Dodder is a parasite of Heather and other plants.

RIGHT: To bee or not ...
A spectacular member of the orchid family, a summer sight to behold is the Bee Orchid. This dastardly deceiver attracts male bees with a part of the flower resembling a female bee. The males land on the fake female, taking the plant's pollen with them.

tle yellow stars that are the leaf rosettes of the carnivorous **Butterwort**. It produces a neat single blue-purple flower. It is the leaves that are of most interest, covered in sticky glandular hairs that trap insects and slowly roll up to digest them.

The last in this botanical terrible trio is **Common Dodder**, or Devil's Guts as it has been known in the past. A parasite on heathers and members of the pea family, it smothers and knits a network of red tendrils (it doesn't need leaves – it doesn't make its own food) over its host. Later in the month look for its innocent-looking pink clusters of flowers.

June bloom
Thousands of plants bedeck our countryside with their reproductive efforts. None are more spectacular than the 27 species of orchids. Spikes of the similar **Common Spotted** and **Heath Spotted Orchids** are popping up everywhere, but you will have to look a little harder for the more subtle insect mimics, such as **Bee** and **Fly Orchids**.

Foxgloves and **Ragged Robin** provide a purple spectacular this month. Both plants turn up commonly in woodlands and hedge banks, but are never as breathtaking as when they occur in large numbers. The burst balloon flowers of the Ragged Robin are best looked for in damp woodlands and meadows. Forests of Foxgloves are best looked for in woodland rides and clearings disturbed two summers ago. This is because the seeds will have had light and warmth to germinate and the time to complete their two-year lifecycle.

Guelder or Elder?
The hedges and neglected corners fill with a heady scent as they 'blow' white with the flowers of **Elder**. These large platforms made from a multitude of tiny off-white flowers are excellent sources of nectar and pollen for many insects. In turn, they attract wasps, which may join in the feast themselves or pounce on the feasters, abducting them, chewing them to a pulp and feeding them to their larvae back at the nest. In fact, over half of my sightings in recent years of our sadly scarce Hornet have been around the blossom of this plant.

Although not as common as Elder, the **Guelder Rose** has similar flower clusters and historically it was referred to as Swamp Elder owing to its preference for damp places. Despite its superficial resemblance to the Elder, the leaves

are not divided into leaflets but look like a currant leaf with three to five points. Also the flowers are made of two kinds, tiny fertile ones in the centre and big showy sterile ones on the fringes.

Whispering grasses

Like an explosion in a paint factory, our un-improved **meadows** and **grasslands**, those small ghosts of what our countryside would have looked like, are transformed into a rich swathes of mingling colour. In centuries past the sight of all these flowers would have really lifted the heart and signalled the arrival of summer proper. Now we have to search hard for torn shreds of this former floral blanket to even get the slightest idea of what it might have been like.

Fortunately for all us romantics, there are still places where this habitat can be enjoyed as well as many nature reserves where you can stumble across commons, rich hedgerows and patches of coastal grassland, which abound with those 'cover girls' of the flower world, orchids, flowering alongside **Yellow Rattle**, **cranesbills**, **Dog Daisy**, **buttercups** and the shredded blooms of Ragged Robin.

If you do find a little bit of grassland, you can soon lose yourself and forget about how diminutive a fragment it is if you sink down to grassroots level. Down low, the buzzing of bi-planes overhead is drowned out by the whirring sewing-machine legs of grasshoppers and crickets. The stems of plants become vertical highways for many insects and bugs, between the ground and flowers and seeds and all the resources that grassland provides. Overhead fly grassland butterflies such as Meadow Brown, Ringlet and Marbled White.

But don't forget the grasses themselves. Roughly 150–160 of these amazing plants can be found in the British Isles. Perfectly suited to our cool, temperate climate, they thrive here. They are also Cow-proof; with the main growth centres close to the ground they can withstand the sort of grazing that is required to create some of our rarest and exuberant June pictures.

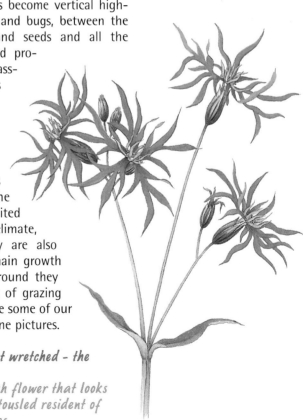

July

This month is one of the warmest, bringing with it the emergence of summer insects. Look out for scuttling Green Tiger Beetles on sun-parched paths and the elaborate mid-air dance of courting Silver-washed Fritillaries. There's a predominance of purple flowers, with Knapweed, Heather and Buddleia at peak. What July has in store for our eyes it denies the ears – many birds fall quiet, their courtship and territorial songs replaced by subdued contact calls or the tuneless begging of newly-fledged young.

ABOVE: Heather on the moor
Heaths and moors really are the place to visit this month. Look out for reptiles and butterflies hanging out in the purple Heather.

RIGHT: Batterpult
To attract bats, use a 'batterpult', a simple catapult, to ping maggots or mealworms (available from some pet or angling shops), into the air. You can feed the bats while watching them hunt!

Mammals

Not all British mammals are hard to see. Those evening Barbecues are good times to observe bats. Our commonest and most widespread is the **Pipistrelle**. It can be recognized in flight as it is very small (Europe's smallest) and has a very random flight. You can watch their aerobatics first-hand by throwing small pebbles in the air. Momentarily they are fooled and swoop towards what they think is an insect, but they soon realize their mistake and change course. Just don't do it near greenhouses! A great way to attract bats without teasing them is to use a 'batterpult'.

See cetaceans
What better way of spending a calm July day than gazing out over a twinkling sea waiting for sights of some of the most awesome mammals in the world, **whales** and **dolphins**. Watching from a headland is often best, especially off the west coast and Dorset although natural bottle necks on the east coast of Scotland are some of the most profitable.

Birds

It's always a pleasure to witness a bird of prey – polished to perfection by millions of years in the evolutionary machine, fine-tuned in the art of aerial skulduggery – make a kill. It's an even bigger pleasure to watch one miss, mess up, stall and give up, a sight guaranteed to make a human feel more at ease with his or her imperfections!

This month tends to be a pretty good one for watching raptors, mainly because most of this year's young have left the nest – but instead of taking to a faultless wing, they end up perched and gauche while still dependent on their parents for food.

Young **Tawny Owls** are often discovered in broad daylight in obvious places looking like awkward, feathered gnomes and at night can be heard 'keewicking' for attention. Young **Common Buzzards**, **Kestrels** and **Sparrowhawks** are easily watched as they sit upon fence posts and trees. They often draw attention to themselves by an incessant wailing for food.

The best sport can be had from watching your local family of **Peregrines**; young will be on the wing and the aerial exploits you can witness will range from food passes between adults and youngsters, plenty of aborted hunting attempts by the young birds, to a range of tutoring by the parent birds as they round up passing birds for their offspring to have a go at.

If you do find a young bird of prey seemingly healthy but out of context, the best policy is to leave it well alone. Young Tawny Owls in particular quite frequently leave the nest before

LEFT: Common Buzzard
Awkward and gauche, young Common Buzzards whine for attention. Buzzards are the commonest large bird of prey in Britain, though not often seen in the east of England.

they can fly but a healthy young owl is a very capable little thing. Many sadly are 'rescued' every year by well meaning but ill-informed people, and end up in already busy wildlife rescue homes. The only exception is if the animal is clearly injured or sick, or at immediate risk, say from dogs or cats, and then it is a simple case of

RIGHT: Tawny Owl chicks
Despite their vulnerable appearance, young Tawny Owls are able to drag themselves up a vertical tree trunk with their talons alone.

What to look out for this month

- Bats in the evening
- Young birds of prey
- Stonechats
- Heathland reptiles
- Solitary bees
- Purple Emperors
- Heather

ABOVE: Heathland's ragamuffin
The Dartford Warbler not only nests among the dense and spiny foliage of gorse, but feeds all year around on invertebrates that also benefit from the plant's shelter: weevils and caterpillars in the seed pods and flowers in summer, and mainly spiders in winter.

gently placing the bird as high out of harm's way as possible. Be careful too, young owls have big, sharp talons!

Warblers in the gorse

The needle nature of gorse leaves means that humans treat gorse with care, but the same cannot be said for the raggamuffin of southern heathland, the **Dartford Warbler**, which nests

right inside gorse bushes. This sombre-looking, long-tailed warbler is rare, but you would expect it to be, for it is insectivorous, with a preference for warm winters, and at the northern edge of its range in England.

Actually seeing Dartford Warblers is more down to luck than planning. However, they should be starting their second brood of the year about now and you have the chance to spy this sneaky bird collecting insects for its chicks. Despite this, many hours can be spent jumping up and down on the spot, both in frustration and in an attempt to see over the scrub, with only the occasional burst of a very debatable song sounding like a scratchy old daisy wheel printer, to keep your hopes up.

While roaming the heaths, keep your eyes and ears open for the 'chat, chat, whoeet' alarm call of the bold sentinel of open places, the **Stonechat**. This bird calls its name, sounding like flint stones being jarred together. It has the refreshing habit of standing in full view on the top of bushes, attracting attention to itself. Not quite so bold but also likely to be seen are **Wheatears**, **Whinchats** and **Yellowhammers**. All will be active, collecting insects to feed their broods or replenishing depleted reserves after the strains of breeding.

RIGHT: Stonechats
Although birds are getting quieter now and do not seem quite so active, the Stonechats are still very visible, being birds of open spaces. Look for family groups.

The scruffy, newly fledged juvenile – watch for feeding and begging behaviour.

Females are duller, but neater.

The unmistakable male in breeding plumage, bold and 'chipper', constantly clacking an alarm at your approach.

Amphibians, Reptiles and Fish

LEFT: Adder
The Adder, Britain's commonest snake, can be sought in rough grassland anywhere on the mainland. They will have dispersed and started hunting by now. Early morning is the best time to catch them out basking.

It may not be the textbook time of year for a spot of reptile rambling, but while planning a day on the heath it is a good idea to try to get there early. You could be lucky enough to catch sight of members of our reptilian fauna catching their first rays of the day.

Most of our reptiles can be found on heaths. It is possible, if you choose your spot, to see all six of our native species. With a little know-how you should be able to obtain glimpses of most of them. **Adders** can be stumbled across almost anywhere and **Slow-worms**, although not really tolerant of dry conditions, may be found where it borders on grassland or woodland.

Hollows in the vegetation provide sheltered hot spots and none are better than the edges of the footpaths that form interconnected webs over most heaths. You will constantly be aware of rapid energetic rustles, just ahead of your footfalls. These are normally created by the ever nervous, highly strung **Common Lizard**. This charming little reptile is rarely observed close up but resist the temptation to continue on your way and simply wait, focus your attentions on and do not shade the spot and in most instances the master of the surreptitious rustle will slowly creep back to its basking stage.

Grass Snakes are laying eggs about now, and you can expect to see them loitering around in damper locations. Because females often hang around their egg-laying site, it is worth checking for them especially where there is a build-up of damp, dead, decomposing vegetation such as piles of old reed stems or the edge of ponds, and of course the compost heap. The warm weather and the decomposition-generated heat provide the clutch of up to 40 leathery skinned eggs with perfect incubation conditions.

BELOW: Grass Snake
Fiddle with a Grass Snake and it will first-ly object by hissing and striking. Catch it and it will ooze an unpleasant smelly fluid from its backside. Persist still further and it will do this - a very alarming and convincing act of playing dead.

Invertebrates

The best places to witness the buzz of insects are Britain's open places, especially heaths.

On a hot, still, sunny day find a thistle- and gorse-free spot, lie down and shut your eyes. At first all you will here is man-made audio clutter, the irritating bi-plane overhead, dogs barking, a car stereo in the car park. But soon this cater-wauling will fade out as closer to your cochlea you will become aware of an incessant droning, not a single tone but many, made by millions of tiny insect wings.

Honey Bees provide the background thrum as they drift from heather to heather, fumbling around for nectar in the tiny almost cerise bauble flowers of Bell Heather and the related deep pink Cross-leaved Heath. Punctuating this domestic insect's monotones are deep rumblings of bumble bees and the higher tones of the smaller, numerous and often overlooked solitary bees and wasps.

At the mention of the word **wasp**, it's a good time to point out that not all wasps are yellow and black jobbies out to get you. You are prob-ably thinking of the eight social species (colonies of sisters founded by the mother) found in this country. The majority of British species are solitary.

They are also one of the heath's best kept secrets and there are many different species fly-ing together this month, but to see them at their best you must seek out their industries.

Concentrate your efforts on firm dry areas of exposed sandy soil, along banks and the edges of paths as it is here you will find the ground perforated with their neat excavations.

These six-legged workaholics can hold you mesmerized for hours: watch for the furtive twitching of Britain's two species of **sand wasp**. These large, leggy black insects (about 2.5 cm long), complete with orange cummerbund, pro-vision their burrows with caterpillars which are paralysed before being stuffed, helpless and

Practical bit

Mirror trick

I always find that animals and plants that live underwater tend to look pretty flat, boring or just plain uncomfortable when out of it. Nowhere is this more obvious than when rock-pooling at the seaside. So instead of turning over stone after stone you can make a natu-ralist's version of a dentist's mirror to allow you to look under overhanging rocks and in crevices. Take a piece of thick but bendable wire and using tough reinforced duct tape (from a DIY shop) attach a travel shaving mir-ror to the end of it. By bending the wire you can achieve the right angle to look in all those previously inaccessible nooks and crannies and even get an idea of what the world looks like from a crab's point of view. Use one to bounce the sun's lights into dark places; it's better than using a torch, although these are useful on dull days too!

undignified, into what will ultimately become their tomb complete with one of the wasp's eggs that will develop to become next year's spindly sand assassins. If only one caterpillar is supplied you are probably watching *Ammophila sabulosa*; if it is more, then it's *A. pubescens*.

Cerceris arenaria is a common black and yellow digger wasp that can be seen performing its sinister miniature airlifts – flying in with the corpses of paralysed weevils slung beneath its undercarriage. These are utilized in much the same way as the sand wasps' caterpillars. Scattered among these different-sized holes are the less tidy spoil heaps of various mining bee species. This term applies to many similar species and identifying them is an entomological nightmare. However, you don't need to get bogged down in their taxonomy to enjoy watching them busy at their daily tasks of digging and foraging.

On a sunny day at the end of the month comes the greatest heathland spectacle of all. Ling fizzes into flower, covering huge tracts of both heath and moorland with its pink-purple froth. Humming frenetically amongst the plants are dumpy bees belonging to many genera. A few minutes watching a nest site will reveal their comings and goings: digging new holes, stealing those of others, aerial dog fights and mating.

A passion for purple

Seeing **Purple Emperor butterflies** is never easy. Even in localities where they are known to be present, they live in low-density populations and can be extremely unpredictable as far as putting in an appearance is concerned.

Their distribution is restricted to Oak woodlands in central-southern England, ranging from Nottinghamshire at the north of their range down to their stronghold in the woods around the Sussex-Surrey border into Hampshire.

It is probably worth limiting your searches to edges, rides and clearings in Oak woodland, during the peak of the flight season, the first two weeks of July. Watch high up in the canopy for the great gliding flights that the very territorial males make as they launch themselves off their leafy throne to see imposing males off their patch, or initiate the aerial seduction of a female that passes through.

So to avoid what Purple Emperor watchers call 'stiff neck', walk the rides and paths of your chosen patch between 10–11.30 am on a warm sunny day, constantly scanning in front of you and you may catch the insect off guard. Any later and they will be setting up and defending their tree-top territories. Male Purple Emperors often visit the woodland floor to catch a few rays or replenish sodium salts lost in the act of sperm production and mating, although they will occasionally visit various animal faeces for this. They are also seen probing the white scurf mark often found around drying puddles.

LEFT: Red-banded Sand Wasp
The Red-banded Sand Wasp is one of the biggest solitary wasps. Look for it fidgeting and fussing along paths on hot sandy heath and moor.

LEFT: Purple Emperor
If you were to score butterflies on a scale of 1 to 10 for magnificence, the Purple Emperor is a grand, but elusive, 11.

Did you know ...?

A putrefying Rabbit? Urine? Or the unsavoury doings of dog or fox? These were baits allegedly used by the Victorian butterfly collecting fraternity to lure the Purple Emperor down from the woodland rafters. Although based on an element of truth (these butterflies rarely visit flowers and seem to need to supplement their diet of aphid honey dew, with salts found in the aforementioned substances) you would get better results and probably have more friends if you simply used a pair of binoculars and a lot of patience.

Butterfly Playgrounds

The Umbra Nature Reserve, Co Londonderry (724355) – Part of the Magilligan dune complex, recognised for its unique succession of sand dunes, the Umbra is open only to Ulster Wildlife Trust members. However, this hidden jewel is home to many butterflies including coastal heath species such as Grayling *Hipparchia semele* and the Dark Green Fritillary *Mesoacidalia aglaja*.

Marford, nr Wrexham (SJ357560) – More than 30 butterfly species come to this site. A disused quarry with cliff faces, woodland and scrub.

Shotover Country Park, Oxfordshire (SP561063) – This has got to be a reserve named after a day watching Purple Emperors, as this is often what they do! It is good for many of other woodland butterflies and a few open space species too.

Bentley Wood, Wiltshire (SU250290) – It wouldn't be going too far to describe this as a bit of a butterfly Mecca. With White Admiral, Purple Hairstreak, Dark Green and Silver Washed Fritillaries to mention a few. The easiest place to see Emperors is along the main East-West ride. Visiting this month you may be lucky enough to see the rare High Brown Fritillary, recorded occasionally in the past. If you do, report it!

LEFT: White Admiral
The White Admiral is top gun of the butterflies on the wing - a strong flier and glider. It will fly quite high, but cannot resist coming down to nectar on Bramble flowers.

The Mens, Sussex (TQ024236) – Rather appropriately this is a Sussex Wildlife Trust reserve, and it is cracking good habitat for the beast that is their logo, the White Admiral. The Purple Emperor is here and often utilizes the larger standard trees for aerial territories.

ABOVE: Purple Hairstreak
The Purple Hairstreak is quite common, but because it spends a lot of its time living high above our heads we rarely see one. Stand back and look at tall Oaks through binoculars - small, fast butterflies zinging around in the lofty limbs are probably Purple Hairstreaks.

Seaton Cliffs, nr Arbroath (NO667416) – Attracts many butterflies to its flowers growing on the sandstone cliffs.

Wellington Country Park, nr Reading (SU724626) – There are a number of habitats, including gravel-pits which are superb for lazy drifts of damselflies. The dense Oak and Beech woodland is sliced up by wide rides from which canopy-gazing for Purple Emperor can be carried out.

Bookham Common, Surrey (TQ121567) – A National Trust property with a rich mixture of different woodland structures and management. Straggly Honeysuckle and Sallow found along the rides, provide caterpillar fodder for White Admiral and Purple Emperor. Mark Oak, car park on the Cobham-Fetcham road and Hundred-pound Bridge are good places to start your quest.

Chiddingfold Forest, Surrey (SU986357) – This is probably the stronghold for that 'wealden Wonder bug' the Purple Emperor. Much of this large tract of Oak woodland has footpaths and other public access. Areas worth visiting are Oaken Wood, a butterfly conservation reserve, and the Woodland Trust's Durfold Wood.

Plants

Not dark and sinister but rich, wet and vibrant, bogs are at their best this month. The luxuriant greens and brick oranges of the 30 or so species of **Sphagnum Moss** provide the juicy centre to these systems and a contrast to surrounding vegetation, which often starts to look a little tired and jaded as age and insects take their toll.

Sphagnum Moss is a unique plant that grows from the top, laying down new growth in the way of leaves and stems and leaving behind the dead, which continue to hold water in their vacuoles long after life has passed. Reach into a clump and pull up a strand and you will see what I mean. The growing tip of the plant is often a vibrant green, orange or red that really brightens up bogs and moorlands.

BELOW: Bog Asphodel
Bog Asphodel is a plant of calcium-poor soggy bogs, hence its old name, Bone Breaker, which refers not to the belief that when sheep consume the leaves their bones become weak but to the overall lack of calcium in the pastures where this plant thrives.

Sphagnum Moss

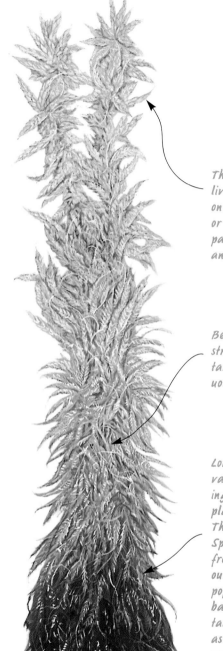

LEFT: The maker of the mire
Without Sphagnum Moss we wouldn't have bogs, moorland or peat!

The top part of the plant is living and can be, depending on the species, green, orange or even bright red. This is the part that photosynthesizes and grows upwards.

Below, the leaves die, but their structure remains. Most important is the big fluid-filled vacuole in the centre of their cells.

Long after the tissues die, the vacuoles retain moisture, soaking it up and holding it in place, so it doesn't drain away. Thus, bogs form. Squeeze Sphagnum in your fist and, as from a sponge, water will pour out. These properties make it popular for lining hanging baskets (not very environmentally friendly or sustainable, as peat bogs are a habitat under threat, much the same as rainforests).

Deep below the surface the leaves finally disintegrate, but they still hold water. This black ooze under a clump of the moss is the beginnings of the formation of peat.

Did you know...?

There is a country saying that when Gorse is in flower, kissing is in season. This, as well as being rather convenient for all romantics, has good biological grounding. Common Gorse flowers all-year round, but the peak flowering is over by July and it is time for the mustard-yellow pin cushions of **Western Gorse**, found mostly west of a line between Dorchester and Edinburgh, and **Dwarf Gorse**, mainly to the east. Both of these species have a much more stunted appearance than their ubiquitous larger relative.

Then as you work your way down the stem into the mush of the bog, the dead material of the plant breaks down slowly, giving way to whites and yellows. Finally, at the bottom, are the dark rich browns, the raw materials that form peat.

All these properties and the fact they can hold water like a sponge long after death mean that these plants are key to creating the unique habitat that is peat bog.

Often associated with these places are **Cotton-grasses** – they can occur in varying densities spotting the landscape with the odd fluffy seed head, or so numerous as to generate a botanical whitewash.

My favourite is the **Bog Asphodel** whose time to shine is now. It starts by producing golden stars of flowers on a spike anything from 5-20 cm tall. Look into the bloom and you have fluffy stamens that look for all the world like miniature Zulu spears. Then as July turns to August they set ablaze the bog as the flowers slowly change to a deep ember orange.

Chat, crackle and pop!

Other than the hum of insects the other characteristic sound of a heath in the stifling midsummer heat is the crackling sound of detonating gorse seed pods. The larger prominent bushes of **Common Gorse** are responsible for this. Since early May these pods have slowly had their supplies of moisture withdrawn, setting up tensions between the two halves of the pod and now they wait in the sun for that moment of relief which sends the seeds scattering.

LEFT: Gorse in bloom
There is rarely a month when gorse is not in bloom. Dwarf Gorse (top) and Western Gorse (bottom) come into their own in July.

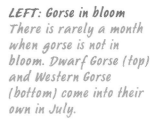

Hot Heaths

ABOVE: Ling or Common Heather
This is the stuff that turns our moors and heaths into a palette of purple this month.

Cannock Chase, Staffordshire (SK005145) – A very accessible bit of heath for all mid-landers. From July onwards a spectacular purple wash as the Ling begins to flower.

Ashdown Forest, Sussex (TQ432324) – As well as being the home to small yellow bears (Winnie-the-Pooh lived here), this gem of mixed wet and dry heath is a top spot for that perfect hazy heathland experience. Classic inhabitants include Adder, Emperor Moth, Nightjar and Dartford Warbler.

New Forest, Hampshire (SU261058) – This vast slab of the county, visited by millions each year, is half shrouded in grazed woodland, but between these areas are many rich boggy bits and patches of dry heath. Most visitors leave having seen no more than the ponies! Look beyond the plentiful roads, car parks and ungulates and you can witness the many more interesting beasts of the heath. It is very good for birds and insects, and you may be lucky enough to see a Hobby. There is also a good 'Reptilery' where you can get your 'eye in' and see the reptiles and amphibians of the New Forest in semi-wild conditions.

The Lizard Peninsula (SW703117) – The heathlands found on Cornwall's Lizard Peninsula are of international wildlife importance, supporting some of Britain's rarest plants. There are several different types of heathland which form a rich patchwork. At North Predannack Downs (SW693167), three main types of Lizard heathland can be found: tall, short and mixed heath.

Studland, Dorset (SZ034836) – Second only to the New Forest for its heathy excellence and accessibility. Get there early as, given the close proximity of a beach and the dead-end road to the ferry, it fills up with cars and people very quickly. On the plus side it's a perfect excuse to do a bit of reptile stalking as all six species are resident.

Thursley Heath, Surrey (SU900399) – The wet heath at Thursley contains rare plants, such as Great Sundew, Bog Hair-grass, Bog Orchid and Brown Beak-sedge. It is also an important site for invertebrates, including the nationally rare White-faced Darter *Leuccorhinia dubia*.

August

This is the month of last minute holiday panics and squeezing the last beach days out of the short British summer. It is also the time of year when the last place you would expect to find any interesting wildlife is on our sandy sea shores. Compared with the rocky shore, golden miles of sand seem barren, but it's just a case of knowing where and how to look.

Mammals

If you find wildlife in the UK dull and pre-dictable and are saving up for a whale-watching holiday in the Azores, you could spend your hard-earned cash on a telescope instead! It's a good investment anyway but never more useful than when you have a pod of **Bottle-nosed Dolphins** or a shark down the end of one!

It is possible to see some 12 species of cetaceans, and the second largest fish in the world, the **Basking Shark**, whilst standing with

ABOVE: Rockpooling
It's the month of a sunny bank holiday and that means beach! Not your usual beachwear, rubber waders are useful for the naturalist who wants to make the most of the tides.

LEFT: Common Dolphin
When using 'scopes and binoculars, keep a wide view as you could be gazing out to the horizon only to have a group of dolphins playing unobserved at the bottom of the cliff.

both feet firmly on British soil. The technique is simple: find a sticky out piece of land that juts out into the sea in the west or north of the country and sit on it!

You increase your chances considerably if you choose a calm, still day. This creates the 'mill-pond' effect on the ocean that cetacean spotters love and of course makes for a thoroughly pleasant day. Even if the beasts don't show, there are loads of plants such as Sea Holly and Thrift in flower, attracting large numbers of insects. You are also in a good place to witness seabird movements or catch a glimpse of the blue bullet form of a Peregrine.

Natural bottlenecks into lochs or estuaries are always worth checking for **dolphins** and **porpoises**. Migrating or feeding fish need to pass through these points, which provide excellent ambush spots, especially if there is a tide ripping through them, creating what is effectively food on a conveyer belt for the marine mammals.

Pondering over the sea requires a fair bit of concentration as it is easy to become hypnotized or daunted by the large volume of water stretched out in front of you. But with a careful combination of scanning with the naked eye and a pair of binoculars, any discrepancy in the smooth ocean surface caused by a shark fin or a cetacean should be picked up.

Look for clues such as a shining flank catching the light or what appears to be a wave breaking, or even heading in the opposite direction of the current, watch the bows of boats for dolphins riding the pressure waves. Large 'snow drift' flocks of seabirds such as Kittiwakes or Gannets often indicate fish near the surface and a good focal point for dolphins or **Harbour Porpoises**.

What to look out for this month

- Whales and Basking Sharks
- Sandwich Terns
- Baby reptiles
- Rock pool worms and molluscs
- Jellyfish

Hot Spots for Seawatching

Bull Point, Co. Antrim (089510) – One of the best spots to see passing dolphins (along with Basking Sharks and whales) is Bull Point on Rathlin Island. The island is accessible by ferry from Ballycastle and is worth a visit simply for its aesthetic value and views across to Scotland, but the chance to marine mammal watch is an added bonus.

Isle Of Man (SC362766) – If you're serious about seeing and learning about Basking Sharks then a trip to the Isle is a must. Visit the Basking Shark Centre in Peel, the base of the Basking Shark Society. This Isle is not bad for cetaceans either!

Cardigan Bay, Wales (SM897410) – Home to a resident population of Bottle-nosed Dolphins. The shore watching is best from places such as Newquay, St. Davids and Strumblehead as well as Mwnt and opposite Cardigan Island. Other species include Harbour Porpoise, Risso's Dolphin and Pilot Whale from autumn into the winter.

Cornwall – There are plenty of cetaceans here, but they tend to be well distributed around the coast. Although a vigil at any headland such as Prawl Point can reward with sightings of Bottle-nosed Dolphin or a Basking Shark, you need to be out on a boat, whether an organized wildlife trip or one of the many ferries to France or Spain or the Scilly Isles. You can see Bottle-nosed, Common, Striped and Risso's Dolphins or even Sperm, Fin, Sei or Hump-backed Whales.

Isle Of Mull (NM607360) – Possibility of Minke Whale, Common, Bottle-nosed and Risso's Dolphins, along with the outside chance of Orca or a Basking Shark. Boat tours led by Sea Life Surveys of Dervaig.

Minch – The bit of sea between the Inner and Outer Hebrides. Greenstone Point, north of Loch Ewe is a good spot to look for Harbour Porpoise, Common Dolphin, Minke Whale and possibly Orca.

Moray Firth, nr Inverness – This has got to be Mecca when it comes to shore watching of the most northerly (with the largest individuals) population of Bottle-nosed Dolphins in the world. Numerous places to watch from including Chanory Point, Cromarty, Burghead and North Kessock. Pop into the Dolphin and Seal Centre, which is off the first lay-by northbound over the Kessock bridge. Hand over your money and listen to the dolphins via hydrophones and with luck watch them in the narrows at the same time!

Durlston Head, South of Swanage, Dorset (SZ031773) – A small group (approximately 10 to 12 individuals) of Bottle-nosed Dolphins are seen occasionally from this point. It is worth a chance as this is a cracking part of the coast with plenty in the way of wildlife interest. The visitor centre at the country park has a hydrophone set up and you can hear dolphins about 80% of the time.

Birds

You may prefer to stretch out on a towel soaking up the sun's rays and watching the gulls against the blue, but if you hear the loud, grating 'keerick' of adult **Sandwich Terns**, it is well worth raising your head for a few moments. These beautiful swallows of the sea are our largest tern and, with their sooty black cap and crest, certainly very handsome. They breed in quiet, sandy coastal locations from May until this month, but as soon as this year's young are able to fly, they wander around our coasts in large family groups.

Even the most crowded tourist resort can be a good spot to watch the adults partake in spectacular water smacking dives for Sand Eels attracted into the warm shallow water. Their spoils are then given up to scruffy, twittering young perched on nearby groins. These adolescents will have hassled the parent birds since they could fly and will continue to do so right down to their African wintering grounds, for which they will depart in the next month.

Other avian exploiters of the less charismatic beach life, can best be viewed on the quieter stretches of our sandy and shingle coasts, or return to the beach when all the day trippers have left. You may well be able to watch them frantically foraging. Many will have young and will be building up reserves for the winter ahead. **Oystercatchers**, hunting for Cockles and other molluscs, or **Turnstones**, working along the necklace of strand debris for Sandhoppers and fly grubs and pupae, are common sights. You may be lucky enough to see smaller waders such as **Ringed Plover** and **Sanderling** running along the edges of the water, picking up small molluscs, worms and shrimps.

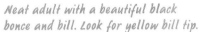

Neat adult with a beautiful black bonce and bill. Look for yellow bill tip.

The vociferous young chase and whinge after their parents. They are much scruffier - no crest and scaly feathers on their back and wings.

BELOW LEFT: Sand Eel
Dinner! It's not easy being a Sand Eel. They are a staple diet of many species of seabird, not to mention predatory fish and the like.

BELOW RIGHT: Clockwork toy birds
Sanderlings do not breed in the UK and will have started their migrations south from the Arctic tundra. They run in and out with the waves, feeding on small invertebrates.

Amphibians, Reptiles and Fish

The embryonic master craftsman has been busy throughout the spring and summer, sculpting what has to be the definition of perfection in August. Compost heaps give up the exquisite 10 cm miracle that is a fresh-from-the-egg **Grass Snake**.

Rough grasslands, heath and woodland rides rustle with miniatures of **Adders**, **Slow-worms** and **Common Lizards**. If you live in the south you may even stumble across youngsters of our two herpetological rarities, the **Sand Lizard** and **Smooth Snake**.

A little tip for reptile watchers eager to see the newest recruits to the reptile world: choose an early morning on a day when warmth from the sun is just making it through, but has not really toasted the ground yet. Reptiles tend to prefer the uneven habitats, tussocks of grass, walls and path edges where there is a combination of open ground that traps heat and provides a basking spot, food (usually small insects), and cover in which to dash when naturalists come stumbling along. Wait silently where you thought you heard the scampering or slithering of a retreating reptile and you will usually be rewarded with a perfect but cautious head, followed by the rest of the body as the reptile returns, more often than not, to its favourite sun lounge.

ABOVE: Young Common Lizard
Common Lizards, also known as Viviparous Lizards, do not lay eggs, but give birth to live young.

BELOW LEFT: Slow-worm: adult and young
Slow-worms give birth to tiny golden replicas of themselves.

BELOW RIGHT: Newly-hatched Grass Snake
These eggs have been developing in the natural incubator of decaying vegetable matter.

Invertebrates

Forget those irritating little bendy plastic buckets and spades – on your visit to the beach take a sturdy full-sized steel fork and instead of a bucket take a shallow tray or small glass observation tank. With a bit of effort you will get to uncover the odd life forms that hide below the surface safe from dehydration, temperature extremes and predation in the damp sand.

Etchings in the sand
It helps to learn to read the etchings in the sand surface as these blow the cover of the animals below and once located the beasts can be harmlessly exhumed with a deft scoop of a fork.

Common are coiled spaghetti-like casts of sand, each paired up with a shallow depression that gives away the presence of **Lugworms** *Arenicola marina*. Small star-shaped marks surrounding a single hole reveal where **Ragworms** *Nereis diversicolor* gingerly reached out to scavenge for food.

Two closely associated holes – one with various patterns of grooves around its mouth – reveal the location of molluscs such as the blushing pink-shelled **Thin Tellin** *Angulus tenuis*. It has paired siphon tubes, one long mobile tube that Hoovers the surface of nutrient-rich sediment that the tide brings with it and another which remains in position whilst water is exhaled through it.

The **Common Cockle** *Cerastoderma edule* is a very common filter-feeding bivalve and most likely to be dug out as it can only live within the top few centimetres of the sand. These dumpy molluscs, which are thought to be ridged so that they stay put in the sand easily, do not look all that athletic, but place one in shallow water and look out for it leaping! Using the protruding muscular foot, it can suddenly lever against the sand surface, resulting in a few rolls, rather less impressive than your imagination may have led you to believe, but of great significance to the Cockle as it allows movement to new feeding grounds or even escape from slow-moving predators.

The master of the sands and the champion digger is the **Pod Razorshell** *Ensis siliqua*. This mighty mollusc is a familiar presence on sandy beaches as its empty shells are well known to any beachcomber, but seeing the live animal in the flesh requires a certain amount of cunning. Walking the lower reaches of the beach, look for shallow depressions in the sand or the tell-tale squirts of water as the streamlined animal pistons into the sand at a staggering half its body length a second! Sprinkle salt on the surface above the burrow, which temporarily irritates the razorshell and it will back up. When just enough is protruding make a quick grab at the shell and pull, if you are quick enough you should have a pod in the hand. If you are too slow the animal will swell the end of its foot in the burrow and will not budge. If this happens, give up and look for another as any further effort could injure the mollusc. Place the Razorshell on the sand and watch as it extrudes its long muscular foot, probes for a while before quickly slipping back under the sand, giving a quick squirt of disapproval as it disappears from sight.

Many animals leave no clues, so even the most astute detective has to resort to old-fashioned trial and error. Enlisting the help of a soil sieve should uncover such oddities as **Heart Urchins**. Look out for the innocent-looking **Necklace Shell** *Polinices catenus*. This is the 'cloak and dagger' operator of the mollusc world, enveloping bivalves in its oversized foot before drilling a neat bezeled hole using acidic secretions and a boring organ.

BELOW: Riddle of the sands
The hole in the tellin shell on the left was the cause of its demise. The innocent-looking Necklace Shell (centre) made the hole and killed the mollusc. You may also come across the 'necklace' egg mass of the killer itself.

ABOVE: Razorshells on the beach
Storms can rip Razorshells out of their burrows and deposit them on the beach. Sadly, all we get to see are the dead animals when this happens. With a bit of nature detective work you can quite easily locate live ones.

ABOVE: Dragonet
While checking out sand pools for invertebrates, you may come across these little fish, which remind me a bit of parrots. They can be very flamboyant, especially the males. You are most likely to see these while snorkelling in shallow water.

BELOW: Jellyfish
It is common to see various spectacular jellyfish washed up in northern waters of the North and Irish seas. Beware - some have stinging cells. Even if the creature looks dead, they can still pack some punch!

Stranded

Strand lines are good places to get clues about life in the surrounding sea, but what you find here are merely sun-bleached carcasses, husks and body parts. To find the living, look for stranded tidal refugees in the equivalent of rock pools. Sand pools of varying size often form on some shallow shelving beaches. At first sight they may seem devoid of life, but free-swimming creatures of sandy bottoms live an exposed existence and are hard to spot. A number of techniques can be used to reveal them. The simplest is to paddle with a shuffling motion to encourage animals to break their cover. Unfortunately, as soon as they stop they are rendered invisible again.

Better still is to enter their world and cut out the difficulties of seeing through the rippled surface. Donning a face mask and snorkel and lying face down is an excellent way of viewing but if you are on a busy beach or with friends this is equal to social suicide, so an old ice-cream tub lid with cling film stretched over a large window cut in it, and floated on the surface of the water is an excellent alternative. Once you have your window into the watery realm the sorts of creatures you can expect to see are the lurkers.

Sand pool community

Common shrimps *Crangon crangon* are usually very common. Watch them shuffle down into the sand and then sweep sand over themselves with long mobile antennae, leaving just the eyes exposed. **Ghost shrimps** *Shistomysis spiritus* are not true shrimps but Mysids, a distantly related group of Crustaceans. They are harder to see as they are transparent, but look out for their coal black eyes. Fish are the most active animals in these pools. You may find small shoals of silvery **Lesser Sand Eels** *Ammodytes tobianus*, bottom-hugging ambush specialists such as young frog-eyed **Dragonets** *Callionymus lyra*, and the infamous **Lesser Weaver** *Echiichthys vipera*.

No fish, no jelly

It looked like an explosion in a marmalade factory, only we were nowhere near civilization of any sort let alone a purveyor of fine conserves. This was the west coast of Scotland one August and the beach was covered in large dollops of orange-brown jelly. Some were obviously fresh, voluptuous and glossy in the sun; others had the look of scraps of deflated balloons, ranging from 40 cm to nearly a metre across. On closer inspection they turned out to be the tattered

and torn wasting bodies of one of the biggest species of jellyfish found in British waters, the **Lion's Mane**. From about this time onwards many species of jellyfish which have spent the summer months growing to their reproductive sizes in the warmer summer seas now congregate in shallow water. All it takes is a rough sea, and they're stranded.

Our commonest species, the **Moon Jellyfish**, is smaller but prettier – recognized by the four purple near-complete circles (its reproductive organs) in its bell.

The other recognizable and often seen species is the **Compass Jellyfish**, so called because of the brown spoked pattern on its bell.

If you do notice large numbers of stranded jellies, check out rock or sand pools, or even the surf if it is calm. Here you may find living specimens gently pulsing. A word of warning – jellyfish sting! Even if they are nearly dead, some cells may have life in them – look out for the stinging cells (nematocysts).

A bundle of barnacles

Depending on your geographical position and the state of the tides you could be in contact with any of around eight British species of these encrusting crustaceans. Despite superficially resembling limpet-like molluscs, they are more like upside-down crabs in their own-made calcium boxes. For a **barnacle** with a dark side check out *Sacculina carcini*. The females of this parasitic barnacle prey on **Common Shore Crabs** – infecting as tiny larvae, before ramifying through the tissue of the host, eventually becoming visible to the human eye as a huge smooth shiny yellow brown mass (not to be mistaken for the granular egg masses of the crabs themselves) on the underside of the unfortunate animal.

Did you know . . . ?

The Moon Jellyfish uses the sun's position in the sky to navigate and congregate to breed. This behaviour leads them often to mass strandings, covering beaches in a blanket of watered-down purple and even making it difficult to swim!

Practical bit

If you find a small stone complete with barnacles, pop it in a jar of sea water and back light it with a strong torch and you should see the flickering feathery limbs as they reach out of their calcified turrets to net tiny food fragments as they drift by.

ABOVE LEFT: Barnacles
Barnacles are not shell-fish as often supposed but a kind of upside-down crustacean (a bit like a crab) in a box. Watch carefully when they are feeding in a clear tank full of sea water and you will see them waving their legs about to catch food.

ABOVE RIGHT: Ragworm
Large Ragworms can give you a nasty nip. Even though they are not active predators, they have a powerful set of black jaws; these are normally hidden in the inverted mouth-parts, but a gentle squeeze on the back of the head will reveal all.

Top Beaches

LEFT: Common Bladderwrack seaweed
The little floats that give this seaweed
its name make a cool popping noise when
squeezed under boot or between fingers.
They serve to lift the fronds up to the
light when the tide is in.

St. Cyrus, Grampian (NO742634) – All the foreshore beasts are here and it has good birds, including a colony of Little Terns, which is wardened in season. A good spot to pick up early migrants too.

Gibraltar Point, South Humberside. (TF556581) – An incredible 400 ha of sand and shingle and all that goes with it, dunes, shingle bars, saltmarsh. Breeding terns, Ringed Plover on the shingle and Short-eared Owl on the dunes plus more sand life than you could shake your fork at.

Ramore Head, Co Londonderry (859411) – Close to a small interpretative centre run by the Environment and Heritage Service of the Department of the Environment, the shore at Ramore Head in Portrush offers the rock rambler much to investigate. The Countryside Centre will help interpret the findings, whilst the nearby Curran Strand (875410) is ideal for beach combing close to where the Atlantic Ocean meets the North Channel.

North Norfolk Coast, Norfolk – Take your pick, just about the entire coast of Norfolk is of interest to the naturalist, with a hotchpotch of sand, shingle, dune and saltmarsh. Check out Holkham (TF892447) and Holme Observatory (TF697438).

South Gower Coast, South Wales (SS417880) – An excellent spot for the time of the year, it has a tendency to get very busy with holiday traffic in season. However, its variety of habitats provide excellent places to find the wildlife that can't run or fly away. As well as all the sandy coast wildlife, this is a good spot for plants.

Exmouth & Dawlish, South Devon (SX980785) – The famed tourist resorts on either side of the Exe estuary are a good place for the family naturalist to visit: plenty of interest for those wishing to partake in the usual beach activities, whilst the wildlife enthusiast can explore the beach and dunes for birds, plants and sand life.

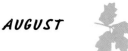

Plants

Purple is a rather popular colour in the floral world this month. Moorlands and heathlands are covered with a 'got to be seen' grand display carpet of pink and purple flowers as the **heather** plants are still blooming. When buying your honey in the supermarket, remember that the bees that feed on heather produce a thick, dark coloured one.

As well as heathers, smaller and more subtle purples and pinks can be found in grassland. **Wild Thyme** *Thymus drucei* and its two relatives, the stronger smelling **Large Thyme** *T. pulegioides* and the **Breckland Thyme** *T. serpyllum* found only on the Breckland heaths, are all in bloom. The rest of this family, the mints, are also at it, providing feasts for the last of the season's insects.

By late August, keep an eye out for the large caps of **Parasol Mushrooms** *Lepiota procera* in fields and at woodland edges. They are edible if you cook them, but be sure not to confuse them with other mushroom and toadstool species. The stalk has an obvious double ring.

ABOVE: Thyme
You often smell this plant before you notice its presence with your eyes. It loves rough grassland, particularly calcium-rich downland, and is a favourite among butterflies.

BELOW: Spring reprise
Look at the Oaks and you will see their second coming of the year, new fresh stems and leaves at the tips of the branches. This is known as Lammas growth.

September

This is a restless, but rewarding, month, with 'last orders' having been called. The last wave of insects emerges and matures; grasslands scratch away with their own rhythm as the grasshoppers and crickets get in their final tune before the frosts; migrant butterflies and moths bustle around the late, nectar-rich flowers. Families go in search of berries; everything that has flowered in the last few months is hanging with seductive seeds and fleshy fruits. Gathering Swallows and martins herald the beginning of the great intercontinental shift as northern-breeding birds head south.

ABOVE: Wild Marjoram
The last generation of summer insects and arriving migrants feast on the late-flowering nectar fonts like Hemp Agrimony, Mint and Wild Marjoram.

RIGHT: Mouse larder in old nest
The bounty of late summer is harvested and hoarded by many, a kind of last-minute supermarket sweep of survival. Here a Wood Mouse has cached his surplus in a disused bird's nest.

Mammals

When Keats famously mentioned autumn and fruitfulness, he surely must have been writing about September. If the summer is the party then this month is the last hour. It's the plants that host this celebration and they are pretty generous too! This month many of our mammals that normally have slightly less fruity tastes, cut down on the invertebrates and jam themselves with super-abundant fruit. Blackberries are a particular favourite of **Badgers** and **Red Foxes** – their droppings as well as being somewhat looser than normal take on a distinct purple hue.

Nature goes nuts

The reason most of us end up getting our Christmas nuts from the greengrocer is obvious this month. **Grey Squirrels** chisel the still-green Hazelnuts in half and the distinctive cases, chipped at one end, are very noticeable underneath the bushes. The squirrels do a very

Be shrewd

Having raced through the summer months in a high-speed orgy of feeding and breeding – pretty much the two things they do best – Britain's shrews are at peak population right now. There can be an awful lot of them too. For example in good habitat the **Common Shrew** can exist at densities of 70 per hectare! This bounty of the micro-mammals is a result of the insect-rich and favourable conditions of the summer months and now they face the winter where many individuals have their short lives (on average 15 months) cut even shorter.

With such a large number of animals dispersing and searching for territories of their own, they are bound to run into other shrews. The high-pitched squeaking frequently heard emanating from long grass or the cover of ivy are the audible testimony that a pair of these solitary hunters have bumped into one another.

With such a flux of naïve shrews on the run it is not surprising that many fall foul of domesticated cats, or simply keel over with exhaustion if they cannot get 90% of their own bodyweight per day in food. It is quite common to find their bodies by paths or on the lawn. Most will be the Common Shrew but do keep your eyes open for our other mainland species, from the minuscule **Pygmy Shrew**, weighing in at 6 g to the gargantuan (for shrews anyway!) **Water Shrew** with its dark fur and white ear tufts.

efficient job in some places of removing every nut from the bush. Many other mammals, including **dormice**, **Wood Mice** and **voles** also join in with the squirrels and now is a great time to have a forage for the fresh feeding signs of each species. This year's evidence, the green, white chippings and nibbled nuts shine out against the dark dank leaf litter, giving you an up-date on the latest small mammal situation in your patch.

LEFT: Grey Squirrel
You may be eyeing up the cobnuts on your local Hazel bushes, thinking they'd be nice to eat. Forget it! Grey Squirrels will get there first. Before the nuts have started to ripen the squirrels will be chiselling and chipping and burying what they cannot eat.

What to look out for this month

- Grey Squirrels
- Shrews
- Birds on passage
- Craneflies
- Brown Hairstreak Butterflies
- Nuts, berries and Hops

LEFT: Pygmy Shrew
Britain's smallest mammal at about 5.5 cm long, the Pygmy Shrew is rarely seen unless brought in dead by the cat.

RIGHT: Common Shrew
Numbers peak around now and lots of squeaking and squealing is going on. Sounds cute to us, but it's a full-blown punch up to a Common Shrew.

Birds

ABOVE: House Martins
Found all over Britain,
although scarcer in the
north, House Martins
are a familiar sight
perched on wires in the
autumn, preparing for
their big migration
southwards.

RIGHT: Bluethroat
This is a spring and
autumn visitor. Look out
for the Bluethroat's
Union Jack waistcoat at
coastal locations across
the country.

From little brown jobs to large white ones, the ornithological world is in turmoil at the moment. Headlands in the south and east are akin to a busy bank holiday weekend at Heathrow airport as the autumn bird migration is in full swing.

Winter visitors flying in, summer breeders checking out, a few species are in transit, landing for a refuel and fuselage check and, to complete the analogy, there are always individuals that are lost, way off their intended course.

Despite the many clues that the autumn bird migration is underway – **Swallows** gathering on wires, the **Spotted Flycatcher** that was always in the garden disappears, and the shrubs echo emptiness, no more calls from **Willow Warblers**, **Whitethroats** and **Chiffchaffs** – it is not as obvious as the spring equivalent. Then the birds are driven by a lustful urgency to set up territories and get a head start on the breeding season, but now the pressure is off and the outflux is a more gradual process.

This is obvious in many of our streets at the moment: some **House Martin nests** lie abandoned, their owners already on their journey; others have their entrances still stuffed full of

the pied 'yippering' heads of the last generation of martins of the year.

Millions of birds that have visited us for the summer breeding season are southbound again, with their offspring of the year. Many that have boarded farther north in places such as Iceland, Greenland and Scandinavia either join us for the winter or use our island like a convenient ornithological motorway services before continuing south to the Mediterranean and Africa.

To behold this season's spectacle the knack is simply to be in the right place at the right time. From now into next month the place to be is on one of the many 'sticky-out' bits. Our headlands, bills, mulls and peninsulas become the focal points for many birds that are moving through, funnelling those moving overland to the shortest oversea jumps and providing a good vantage point to spy seabirds passing by offshore.

If you don't live near moorland and fancy seeing a **Ring Ouzel**, or, if you fancy learning all your warblers in one day, are turned on by rarities like a **Red-breasted Flycatcher**, or simply wish to witness the spectacle of birds such as Swallows doing what they are famous for, the time to do it is now! Choose your day carefully and according to the weather. For the best results think like a bird that has to conserve its resources. Early mornings on days with an inshore wind are best as with these conditions any birds leaving land are unlikely to proceed, and 'bunch up' waiting for more favourable weather. These conditions are also likely to help any birds travelling the opposite way and heading for land.

Top Proms

Holyhead, Wales (SH228833) – From this month onwards our native populations of birds such as **Chaffinches** are swollen by visitors from Europe. A trip to Holyhead and nearby South Stack Cliffs nature reserve is well worth the visit to see autumn migrants.

West Light View Point, Co Antrim (092 517) – Part of a RSPB reserve on the west of Rathlin off the North Coast of Ulster, Guillemot, Kittiwake, Razorbill and Fulmar can be seen on the cliffs, whilst Puffins nest in their thousands. You may also see Manx Shearwater, skuas and Gannets. The island accessible by ferry from Ballycastle.

St. David's Head, Dyfed (SM734272) – This part of the coast runs east to Strumblehead and forms the most westerly bit of Wales. It provides a good spot for watching the activities of passing seabirds and many small birds can also be seen as they move down the coast. **Chough** still haunt the cliff tops and are worth looking out for.

Scilly Isles (SV905105) – This archipelago is a worthwhile adventure, well known among the birding fraternity for both very rare transatlantic migrants as well as more common European birds.

Flamborough Head, Humberside (TA225705) – Sticking into the North Sea there's masses of seabird activity. Sit in the late summer chalkland flowers, focus your binoculars on the water and you'll see many birds passing by. Good for **divers**, **Gannets**, **shearwaters** and on a day with breeze from the sea, queuing songbirds waiting to escape to the Continent.

Spurn Head, Humberside (TA417151) – This sand and shingle spine of coast is one of the best places for migrant birds in Europe. Famous for large 'falls' of small birds as well as **waders** and **terns**, the latter sometimes harassed by a visiting **Arctic Skua**.

Snettisham, Norfolk (TF648320) – Waders are also on the move. Nowhere is this more spectacular than the Wash, and this is best witnessed from Snettisham. Look for **Bar-tailed Godwits** still in the remnants of pink breeding flush; some **Dunlin** will still have black bellies. Find a **Little Stint** and listen for the whistling **Whimbrel**.

Cley, Norfolk (TG053440) – You never know what you may see at Cley. The whole of the North Norfolk coast is just twittering, and every bush seems to have far more birds in it than it should. With **Nutcracker**, **Bluethroat**, **Great Grey Shrike** and **Rosefinch** all possibilities, it's a phenomenal bird list, and you may just be lucky.

Isle of Portland, Dorset (SY668755) – Being a small area on which to focus your attentions, it is a truly excellent place to visit for interesting passage migrants and a prime slice of the seawatching experience, with over 300 species of bird having been recorded here.

Beachy Head, East Sussex (TV586956) – Look out to sea and you often see the passage of terns and Gannets, while around your head wheel the last views of this year's Swallows and martins. The scrub on the top provides temporary shelter for many summering warblers. Look out for **Blackcaps**, **Garden Warblers**, **Chiffchaffs** along with **Goldcrests** and **Firecrests**. Other migrants such as **Ring Ouzels**, **Redstarts** and **Pied Flycatchers**, late **Cuckoo** chicks and Hobbies frequently rest here too.

Non-breeding plumage

Breeding plumage

LEFT: Bar-tailed Godwits Snettisham in Norfolk is a good place to see these spectacular birds at this time of year.

Our native populations of birds such as **Chaffinches** are swollen by visitors from Europe. They end up on the west coast after skipping from field to field continuing in the direction they set off, eventually accumulating on headlands such as Holyhead before making the short skip over to Ireland. Chaffinches may not seem the most interesting fodder for your binoculars, but they are very obvious and when in large flocks are spectacular visual clues to the bird movements going on at present.

LEFT: Chaffinch
Chaffinches on their own might not generate much excitement but look out for spectacular large flocks of these birds moving around. Many visit from the continent, particularly if the weather there is bad or if food resources are poor.

Amphibians, Reptiles and Fish

Unlike frogs and toads, **newts** tend to hibernate only a short distance from ponds and streams and will start to leave the water from September onwards. They will look for a suitable hiding place to spend the winter – under a log or stone. Keep an eye out for them after dark near your pond, especially on wet nights.

If you want to encourage newts to hibernate in your garden, keep areas of rough and long vegetation, especially around your pond, to give newts somewhere to hide and shelter in. Leave piles of leaves, logs or even rubble near your pond. A compost heap provides an important source of shelter and food.

RIGHT: Nocturnal newts
If you want to try and see newts heading off to hibernate, go out after dark on a wet night and wait close to a pond or stream.

Invertebrates

With a lot of the natural world slowly bowing before the curtain of autumn drops, it is good to know that some summer activities are still going strong. With nothing to lose, many of the **grasshoppers** and **crickets** are still scratching out their grassroots symphonies in September and some continue to do so in sheltered places well into October.

One dainty member of the cricket clan that is often hard to spot becomes slightly less reclusive this month. The **Oak Bush Cricket** is a small light green cricket, which spends its entire life more often than not in Oak trees, even laying its eggs in cracks and fissures in the bark. Entirely nocturnal, it lurks on the underside of leaves during the day, becoming active at night.

It is the nightly wanderings of this insect that often lead it into our lives. Like many insects it is attracted to lights, and those lucky enough to live close to an oak tree stand a good chance of waking to find one of these emerald orthopterans dangling from their lampshade.

Nobody really knows why they turn up more frequently now; it's possible that they may be making last minute attempts to find food, mates or are simply dislodged by falling autumn leaves or the stronger September breezes.

Male **Dark Bush Crickets** throw their well-spaced chirps from the safety of shrubs and bramble patches. If you want to find one of these, as an alternative to the *Cricket tricks* (below), apply brute force and beat them out during daylight by vigorously shaking low branches over a white sheet. By shaking branches you may turn up the smallest of the British crickets, the delicately freckled and wingless **Speckled Bush Cricket**, as well as a couple of common but outrageously bedecked **moth caterpillars**.

ABOVE: Oak Bush Cricket
For some reason Oak Bush Crickets seem to be attracted to lights and because of this they often enter our houses around about now.

BELOW: Speckled Bush Cricket
These are the last chirpers and lend their lazy chirrups to damp dew-soaked evenings.

Practical bit

Cricket tricks

If you hope for a visit from an Oak Bush Cricket, here are a couple of tricks that may help you get a glimpse. The most obvious is taking light to a tree. Either a mains-powered cable or service lamp or even a powerful torch with long-lasting batteries strapped to the trunk of a tree can lure the beasts in and while you are waiting you can listen out for the song of the male. This is odd in that it is not produced by mechanisms on the wings; instead these insects engage in some fancy footwork, drumming their 'trr trr trrrr' song out on leaves with their hind legs.

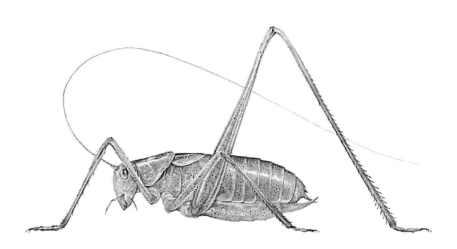

ABOVE: Pale Tussock Moth caterpillar
The Pale Tussock Moth caterpillar is a yellow and green toothbrush affair. Along with the equally punk Sycamore Moth and Alder Moth (Acronicta aceris and A. alni, which, despite their names, may also turn up on Oak) it will soon be taking its last mouthfuls of the season before retiring to the cocoon to pass the winter.

The Daddy Long Legs phenomenon

It's a very British thing. If it's gangly and has long legs then it falls immediately into this category, but what exactly is a Daddy Long Legs? Depending on who you are and what you've seen, it could be a spider, one of a curious bunch of Arachnids known as **Harvestmen**, or more than likely, if it has wings it is a fly, to be precise a **Cranefly**.

September really wouldn't be the same without one of these clumsy characters knocking into your windows or hanging around your porch light. There are numerous different species at wing throughout the year, but this month sees the mass hatching of one of the larger and commoner species, *Tipula padulosa*. If you do not manage to see it attracted to lights, go for a walk in damp long grass and you will not fail to spot the bumbling, battering flights of these less than proficient aeronauts.

Looking somewhat like a cluster of bent hat pins, their shape does not seem to lend itself to winging gracefully through the grass stems as it is often their crashing and buzzing which gives them away as they ricochet from the vegetation.

Check out damp areas of short plant growth, such as the mossy edges of a stream or even the damp end of your lawn and you may well come across evidence of the other side of their life. Like all flies, they have a larval or maggot stage. The last of these is a dull grey beast with very thick skin, known as a 'leatherjacket' and this is the main consuming part of the lifecycle as the adult flies rarely eat. After spending many months of the year under the soil feeding on detritus, roots and occasional plant material dragged down from the surface, they pupate. The black/brown remains of these crunchy casks can often be found poking half out of the soil.

LEFT: Bobbing gnat
You may notice adult Craneflies engaging in a bit of behaviour that has given them the colloquial name 'bobbing gnats'. The females (identified by their sharp-ended abdomen as opposed to the club-like tip of the male) can be seen throwing themselves repeatedly at the ground - this curious bouncing act is actually egg laying, getting up enough momentum to prick the surface with her dart-like ovipositor, depositing the eggs that will become next autumn's batch.

LEFT: *Alien invader?*
Looking like something that would have inspired H.G. Wells, Harvestmen are easily recognized by their stilt-legged gait and small round body (not separated into two parts like spiders) held clear of the ground by the cradle of legs.

BELOW: *Dor beetles*
These trundling armoured waste disposal units can be found on paths or flying to lights on warm evenings. Often their presence is noticeable when their skeletons and wing cases turn the droppings of fox and Badger into glittering, metallic-purple jewels.

Common Dor Beetle

Horned Dung Beetle

The other Daddy Long Legs are both Arachnids, the **Daddy Long Legs Spider** *Pholcus phalangioides* is that very spindly affair often found in un-dusted corners of houses or sheds. Try and touch it and it blurs into invisibility by vibrating itself in its web so fast it is hard to focus on.

The Harvestmen (Opiliones) are the last bunch to get branded with the name. We have in the order of 23 species of this odd-looking scavenger. Some have long legs like the common *Leiobunum rotundum*; others have short legs. They spend most of their life skulking around in damp places and they often find their way into canvas tents when you are camping. They get their name from the fact that they are more noticeable now as most have been maturing throughout the summer.

Dung beetles, as their name suggests, live on animal dung (mainly that of herbivores such as rabbits, sheep, cattle, elephants, etc.) – they feed on dung in both the adult and larval stage. Some dung beetles simply live and breed in the dung heaps left by animals, but others bury the dung in some way or other before eating it or laying their eggs.

Two dung beetles, the **Common Dor Beetle** and the **Horned Dung Beetle** can be seen, on warm evenings, flying over cattle fields and

male

female

LEFT: *Minotaur Beetles*
Whereas the similar Dor Beetle feeds on cow dung, the Minotaur feeds on Rabbit droppings. Note the male has bigger 'horns' than the female.

settling on cow dung. Most of the dung-burying species excavate tunnels under dung heaps and then haul down bundles of dung into these underground chambers where the adult beetles feed and lay their eggs. This scavenging activity provides a useful service by removing the dung from the soil surface and hastening its breakdown in the soil.

Pennies from heaven

The flight of a female **Brown Hairstreak** has been described as being reminiscent of a gold coin tumbling from the sky and it was because of this that one of its old names was the Golden Hairstreak. This gorgeously understated denizen of wood edge and Blackthorn scrub is one of the latest butterflies to emerge. It is normally on the wing from July, through August and in early September you stand the best chance of seeing one as it is the peak time for egg laying.

This butterfly seems to be the kind of insect bumped into and seems to me to resist deliberate seeking! Look out for it in wooded regions of Devon, Dorset, the Weald and a few localities in Hampshire, Gloucestershire and Oxfordshire. You can increase your chances by looking only in the warmest weather – it rarely flies when the air temperature is less than 20 °C. This butterfly is late to rise and early to bed, active between mid-morning and afternoon.

It is usually the female that is most commonly seen, when she descends to lay her eggs on the bark of Blackthorn twigs, usually on the junction between the latest growth and last season's below head height. The eggs she lays now will hatch in the following spring. The larvae will feed on the leaves of the Blackthorn. She may also be seen sipping from hedgerow flowers – such as a late bramble or thistle.

The dowdy males spend their life at the tops of the highest 'standard' trees where they have a genetically pre-programmed rendezvous with the females. They remain here fuelling their short-lived passion on aphid honeydew.

Plants

Even the plants are at it! Hit the hedgerows and it is very evident. Everywhere, berries, pods, haws and hips are letting it all hang out!

This is the time of the year where fruits flirt and flaunt their edibility, seducing the taste buds of many animals. A visual favourite that is poisonous to us, but not to those for whom the scarlet glossy berries are advertised, is **White Bryony**, the only native member of the cucumber family, whose true lineage is given away by its very cucumbery pale green, lobed leaves and corkscrewing tendrils. Others to look out for are the nightshades: the commonest is the **Red-berried Woody Nightshade** or **Bitter Nightshade**. This is often mistaken for **Deadly Nightshade**, which is more restricted to chalk soils and bears larger single black marble-like berries, that sit in a cup created by the fleshy five-starred sepals.

It seems like a one-sided relationship, with the harvesting animals benefiting the most, but when surrounded by a desirable fruit it is amazing where the seeds can travel. Many white laundry days are ruined by the purple squits of Starlings, Blackbirds and thrushes that are currently squabbling over the heavy fruiting heads of **Elder**. This can provide a good focal point for close-up birdwatching, but return at night, armed with a torch (a red-filtered one is best) to a good bush with open areas beneath it and you

may catch glimpses of voles and mice helping themselves to those fruits knocked off by the birds. Many of our small mammals can become surprisingly tame when faced with the 'stuff face or run' option as they are desperately laying down fat reserves and stashing what they can. The mammal watcher is helped by the fact that mammal numbers are at an annual peak and it is not uncommon to see a Field Vole in broad daylight whistling up and down bramble ladders to pinch the best **Blackberries**.

Get berried!

At the moment both leaves and fruit are present, making it a good time to pick out various species. There are plenty and, I hasten to add, not all as edible to us as to the animals they're targeting. But **Sloes**, the fruits of all that fizzy flowering **Blackthorn** of April, should abound and you may well stumble across the black bundles of berries of **Purging Buckthorn** too. If in woodland or in the higher areas you will not fail

ABOVE: Blue Tit and Elderberries
With the bounty of harvest time many birds switch their diet from insects and the like to seeds. Here a Blue Tit, well known for its eclectic tastes, is eyeing up some Elderberries.

Practical bit

Boot Garden

A good excuse to get muddy in the name of science! You know how plants seem to get everywhere, any crack and crevice in a pavement or patio if left undisturbed will sprout green stuff. Have you ever wondered how they get there? Well, after your next walk, scrape off the mud from the bottom of your boots, mix it into some water. Take a seed tray full of sterile soil (bake it in the over to kill any living seeds that might be in it) and pour on your welly boot solution. Cover with cling film and leave in a light place for a week or so and you should start to see an utterly amazing selection of little plants germinating. You would never have imagined you had all that life hidden in the tread of your boots!

RIGHT: Rosehips
These wild rosehips are a great favourite of finches and small mammals. I've seen Bank Voles scale their thorny heights, risking being spotted by predators, just for a mouthful of rosehip.

BELOW: Rowan berries
A colourful tree in any season, the Rowan's late summer display of burning red berries can be a sight to behold, assuming you get a chance to behold it before the birds move in and scoff the lot!

to notice that otherwise rather introverted little tree, the **Rowan**. Elegantly flowering in May, it gets drowned out by lots of similarly flowering plants and the abundance of other spring spectacles, but now it takes the spotlight – a blaze of bountiful clusters of crimson berries, so popular with the thrushes. Towards the end of the month its leaves join in too, one of the few British trees whose leaves turn red.

My favourite fruit belongs to the inconspicuous **Spindle** tree. A plant of hedgerows and light woods, it never really stands out but seek it this month. The matt red four-lobed seed pods, which when ripe split to expose a vivid orange-coloured seed, are a hidden gem well worth the search.

Soul food?

There are some intriguing myths about the bramble bush. One tall story is that when the Devil was cast out of Heaven, he landed in a prickly bramble bush. It is believed that it is unlucky to pick blackberries after September 29th (St. Michaelmas Day) because the Devil curses them – he may take your soul.

Nevertheless, it's good for the soul, in these days of non-seasonal supermarket shopping to nip out for a forage. It is also worth putting some of your excess sloes, **rosehips** and blackberries in your freezer ready to bring out as a welcome supplement to the bird table this winter.

The botany of beer

This is a plant that is of interest, even to those with the most reluctant botanical tendencies.

LEFT: *Blackberries*
It's the berry on the end of the fruiting shoot that ripens first and is the juiciest. The others then ripen in turn but do not quite live up to the first.

It may come as a surprise to those who drink that elixir known as beer, that **Hops** are not just found in a glass but the plant is actually native to the hedges and woodlands of the south of England, where it sprawls and drags itself up other vegetation with its tiny hooked stems.

So rather than make do with the dry, dusty adornment of your local pub bar, why not look it up in the wild? Because it is so familiar, the first things you notice about this plant during September are the mature and scaly female fruiting heads, looking a bit like diminutive pineapples (the bit that is usually harvested to flavour beer). Hops are what is known as dioecious, which means that there are both male and female plants – the male plant flowers earlier on in the year.

So what is it about these strange fruits, that makes them flavouring for beer? Take a fruit and peel back the scales and at the bases are yellowish glands which produce a mixture of aromatic oils, the magic substance that makes 'bitter' bitter.

But despite its popularity now, it was a while catching on in England. On the continent Hops were the flavour back in the 9th Century but it wasn't until the 15th Century that we used them in this country. Prior to that we were using **Bog-myrtle**, **Wormwood** and **Ground Ivy**. But the drink kept better when flavoured by Hops and eventually it became the favourite. It was then that Hop farming took off and became profitable.

BELOW: *Hops*
The long tangles of this famously tasting plant can be found in hedges as a native, but it has also been distributed for commercial cultivation and for its ornamental looks.

The male flowers were open at their best during July and August and are found on a different plant from the female flowers.

The female flowers are hidden between the large papery scales that become the fruiting body ultimately to be found adorning the bars of pubs.

Did you know ...?

Ever wondered why blackberries are so variable in taste, texture and size? It's because there are well over 2,000 micro-species of this common plant.

October

All the other seasons span several months, but autumn, or at least the great seasonal test card of leaf change, fits perfectly into October. There is an urgency to October – in our rivers Salmon push on upstream during the day, Sea Trout at night. Mainly unnoticed, birds continue to mill about, the summer breeding migrants having left and the arrival of the bulk of winter thrushes, waders and waterfowl all imminent. Small mammals are storing food, or eating as much as they can before they hibernate.

ABOVE: Autumn leaves
There's nothing more evocative than the changing leaf colours of autumn. Head down to the woods and revel in the vibrant oranges, reds and yellows.

RIGHT: Red Deer
The Red Deer is our largest native land mammal and one of six species of deer that occur in the wild in Britain.

Mammals

I have to shout about this one every October! It is unmissable! The cracking of antler against antler is a sound that has been echoing through the woods and hills of Britain for thousands of years. The **Red Deer** rut is upon us again, and here's the best bit; this behaviour is not so subtle as to be unnoticeable to anyone that hasn't got a PhD in mammal behaviour. Red deer stags are extroverts – bellowing, bush bashing, clashing and stinking their way through October.

There are also loads of places to go to see them in action, from Scottish highlands and places such as Exmoor to deer parks, some of which can be found very close to most urban centres. Richmond Park is one of these places: the scene before me is very surreal, against a modern cityscape on the edge of a Sweet Chestnut and Birch copse glowing golden in the sun. The monarch of the glen, looking strikingly primeval is belching out his challenge to other equally well adorned stags in the area, shreds of bracken hanging limp around his brow. Not a heather moor to be seen, just a passing Jack

What to look out for this month

- ■ **Deer rutting**
- ■ **Tawny Owls hooting**
- ■ **Ground beetles**
- ■ **Spiders spinning**
- ■ **Mushrooms and toadstools**
- ■ **Leaves changing colour**

Russell and owner, and a lady with a pushchair, and in place of where you would expect a Golden Eagle a trio of squawking Ring-necked Parakeets head for their roost.

It may seem like a bit of a cheat, and yes a trip into the hills of Exmoor or the glens of Scotland would be very romantic, but if you cannot manage this (and most of us can't), a trip to your local deer park is just as good and in many ways better because you get unbelievably close to the animals and they will carry on behaving naturally. October is the time of the rut, when places where Red Deer reside echo to what sounds like numerous reluctant Harley Davidsons being kick-started.

The male deer or stags have been waiting all year for this. Their antlers, grown from the bony coronets on the forehead since the spring are now in their full splendour. The animals have put on a lot of weight, particularly around the neck and head. With the skin and hair of the head and neck thickening and the muscle becoming a third thicker in girth, each stag has become a living battering ram. They are dedicated, enslaved by their hormones for the duration of the rut, hardly feeding at all, constantly displaying, bellowing, tree bashing, scraping and wallowing, letting neighbouring stags know they mean business while at the same time trying to round up a harem of does.

You can see all this behaviour in the real wild, but I have a friend who spent nearly an entire October up a tree dressed in camouflage clothing to get a photograph of Red Deer wallowing. Nip down to your local deer park, like Richmond, and you could have filmed it in an afternoon and had time to fit in a bit of early Christmas shopping!

A word of advice: be aware that stags are big, they have testosterone goggles on and will not be all that chuffed if you get between them and their does or cover. So enjoy the big autumn spectacular and keep a bit of distance and the dog on a lead.

Deer Hotspots

Aviemore, Inverness-shire (NH967095) – For the real thing check out Ryvoan pass. Our two native species, Roe and Red Deer, hang out in native Caledonian pine forest. This is also a site for Wild Cat although for most people this is purely academic as they are virtually impossible to find.

ABOVE: Chinese Water Deer
The smallest deer living wild in Britain along with the Muntjac.

Cannock Chase, Staffordshire (SK005145) – This relict piece of heathland, in the middle of the Midlands has five deer species to be seen on the heath at first and last light. With luck you can notch up Red, Roe, Fallow, Sika and Muntjac.

Thetford Forest Park, Norfolk (TL830830) – A mix of habitat and cover provide a good accessible site for Roe, Fallow and Muntjac.

Woburn Park, Bedfordshire (SP9734) – Thanks to the activities of the 11th Duke of Bedford, the park and its surrounding countryside teem with deer, not only Fallow and Red but the two smaller exotic Chinese Water Deer and Muntjac, introduced by the Duke at the turn of the century, and Père David's Deer. Tel: 01525 290666.

Wilmersham Common, Exmoor (SS8743) – A truly primeval experience best visited just before dawn or dusk when large harems of Red Deer each with attendant stags, may be seen. It is also when they are most vocal. To hear the gruff bellows of the stags reverberating around the coombes, especially on a fresh misty morning, is simply intoxicating.

Duryard Valley Park, Exeter (SX922966) – A dawn or dusk visit to the specially made deer hide can reward you with excellent views of Roe Deer all year around.

Richmond Park, Surrey (TQ1972) – The ultimate deer-watch experience! Both Red and Fallow are kept in this large and lovely urban location. Best on a bike as it is a big place and it could come in useful for a hasty retreat as the males, having lost their fear of humans, can be a little aggressive.

Birds

Being nocturnal, well-camouflaged and woodland birds all add to the problems of observing a wild owl. But our largest owl, the **Tawny Owl**, is fairly 'birder' friendly as far as owls go. For a start it is an adaptable bird and tolerant of human disturbance – the only owl to have populations in our cities. In fact, some of my best 'owling' moments have been in a car park in Bristol, watching a bird hunt young rats foraging around a litter bin. In addition, as Tawnies become territorial they can be encouraged to come to you. They are setting up territories at this time of year, pairing off and searching out nesting sites. The males will bring their females food as part of their courtship ritual.

During October there are a few things in our favour. For a start the leaves are falling, allowing better views. Also there is a lot of shouting in the owl world after dark: the young have been ousted from the territories in which they were raised and they and other birds are now competing for territories, which must supply not only food through the winter but also have suitable resources such as roosts and nest holes for the breeding season.

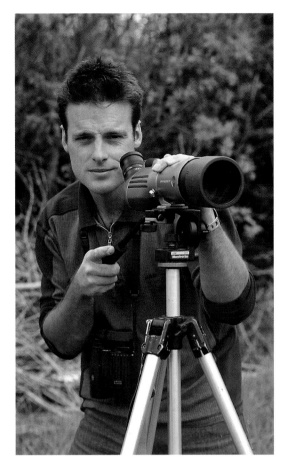

LEFT: Birdwatching
Although a telescope is not essential it does add another dimension to birdwatching. Binoculars, on the other hand, are essential, so you simply must get a pair if you haven't already!

LEFT: Tawny Owls
The Tawny Owl is probably one of the easiest owls to see, especially at this time of the year when it is getting territorial again and responds well to impersonations.

Practical bit

Put your thumbs together and blow!
Try imitating a Tawny Owl's hoot by blowing between your thumbs into an airtight cavity produced by your cupped hands. A bad owl impression can be generated, often enough to get the blood of a resident male boiling! Do this near where you hear owls calling and it is almost guaranteed that you can see an owl by torchlight.

Play fair though – once you have set eyeball on your bird, leave off the impressions and let the owl return to its business.

Amphibians, Reptiles and Fish

River fish take advantage of the higher water levels this month. Autumn spates make getting upstream and over waterfalls and weirs easier. On many rivers the spawning run has started. **Salmon** travel during the day and **Sea Trout** at night to reach their spawning stretches or redds.

The Sea Trout breeding season is usually during early winter, when the adults pair up in small, gravel-bedded streams. After a brief courtship the female cuts into the gravel with slashes of her tail to form a redd to hold the eggs. She sheds a stream of round pink eggs into the redd; the male then moves forward and ejects a stream of milt, which contains the sperm that fertilizes the eggs.

LEFT: Sea Trout
It makes sense if you are a fish to wait until the first real big rains of autumn. This raises the water level, making it easier for fish to start migrating up their spawning rivers.

Invertebrates

Keep an eye on the weather and the tides of your favourite beach and get out there after a storm. The strand line is where it is at and these are easiest to identify on smooth beaches comprised of sand or shingle. Rocky beaches will also reveal a wonderment of wildlife. An eyes-down walk should reveal the overspill of oceanic life – the bubblewrap-like egg masses of **Common** **Whelk**, **Cuttlefish bones** and storm-torn **Bladderwrack Seaweed**.

Look out for floats of **By-the-wind Sailor** *Velella velella*, a sea-drifting relative of jellyfish. Where these are present you can sometimes find the large globular purple-suffused shells of its specialist predator the **Violet Sea Snail** *Janthina janthina*.

RIGHT: Bladderwrack Seaweed
This seaweed grows on rocks and breakwaters on the middle shore, but in a storm they will wash up on the strand line of a beach.

Goose bumps
Other weirdos of the strand line at this time of year are the **Goose Barnacles**. These normally attach themselves to floating wooden flotsam and drift the high seas beneath their inverted surfboards. You are most likely to see the **Common Goose Barnacles** *Lepas anatifera*. These odd creatures, once thought to be the early life-cycle stages of the Barnacle Goose, are like the **Acorn Barnacles** of the shore crustaceans but they have that well-kept

look and a small resemblance to the neck and head of a black and white goose. A closely related species is the **Buoy Barnacle** *Lepas fascicularis*, which has a shorter stalk, duller plates and manufactures its own buoyancy aid.

Mice that swim

A favourite and easily overlooked creature, which normally spends its life scavenging in muddy sands, is the **Sea Mouse** *Aphrodite aculeata* – not a mammal but a rather strange-looking and massive (up to 15 cm long and 5 cm wide) scale-worm.

Adaptations to ploughing through mud have stretched the worm body design to its most alluring. When beached, to the untrained eye, it looks like a drowned balding dead mammal or worse. But pop *Aphrodite* into a container of sea water and providing it hasn't been beached for too long, the dull brown hairs that fringe the body transform into a shimmering, psychedelic and iridescent garb. A healthy specimen may even start to move around using the many bristly parapodia along the side of its body, and a pulsating motion may be evident as water is pumped through a cavity hidden beneath the scales on its back. These are not visible as they are shrouded by the hairs and bristles which prevent fine muds and silts from blocking up this elaborate breathing system.

Spiders' webs

The flossy trimmings of spider webs in our hedgerows, though associated with the autumn, have been present most of the year. It is just that now the autumn mists micro-bead the strands with dew, making them easy to detect.

The **orb web spiders**, those that spin the classic circular webs, are the most noticeable and of these two of the largest common species are the **Garden Orb** *Araneus diadematus* and *Nuctenea umbratica*, which is a lot darker, slightly flattened and has six indentations on its abdomen giving it the appearance of a studded Chesterfield sofa. The webs of both can be found in hedgerows.

Most of the males have died by now but the larger and spectacular pea-sized females are still common. When a web has been found, look for the spider resting in a shelter somewhere to the edge of the web. Although nocturnal, they

ABOVE: Goose Barnacles
October gales can throw up some living flotsam in the form of this curious crustacean, the Goose Barnacle.

BELOW: Sea Mouse
Although it looks like a mammal, the Sea Mouse is the most wonderful of worms.

Practical bit

Weaving wonders

Most **spiders** rebuild their webs nightly, so to watch this you must return after dark. If this seems a little adventurous, catch them out early in the morning, when they are often to be seen resting in the middle of the web. Make a loop of wire slightly smaller than the web, place behind and gently push through. This, if done gently, will transfer spider and web and providing a hiding place is provided, the whole caboodle can be set up at home. With a little luck you can witness these master weavers in the comfort of your own home.

The Garden Orb Spider is also known as the Garden Cross Spider owing to the pretty and distinctive white cross-shaped markings on its abdomen.

RIGHT: *Marble Galls*
These little life capsules are found on Oak trees and each contains the tiny grub of a gall wasp. Whether the grub is in or out depends on whether there is a hole in the gall's surface.

are easily lured out by trapped prey, a tuning fork or a gentle tickling of the web with the end of a piece of grass.

The evidence of many young spiders and adults of the smaller 'money' spiders or *Linyphilidae* can also be seen on these spider dawns. The grass can often appear silver with many single straight lines of silk caused by millions of minute spiders trying to hitch a lift on the breeze, a random form of travel known as 'ballooning'. This can often be replicated on the tip of the finger. By gently blowing on one of these small spiders it lifts its body in the air, facing the breeze, and produces a strand of silk

which eventually generates enough lift to send the creature skywards; the threads on the lawn are the results of millions of spiders having landed or attempted take off.

How galling!

Once the leaves have blown off the trees, many details that remain hidden in the green swathes for much of the year are revealed – birds' nests, squirrel dreys and a multitude of different and strange-looking **galls**.

Galls are mainly the products of the activities of a group of minute insects known as gall wasps.

The tiny insects' egg is laid in the tissues of the tree. As the grub develops it distorts the normal growth, creating a micro-grub utopia, and providing a buffer against the extremities of climate, and food resources to fuel its development.

The most familiar is the smooth, spherical marble gall found on Oaks. These are produced by the asexual generation of a tiny **wasp** called *Andricus kollari*, many of which will have emerged, a tale told by a neat round hole on the surface. Find one without this and seal it in a clear bag and over the next few weeks the tiddly insect may emerge.

From now on into the winter some of the most numerous galls become tit and finch fodder. Woods can instantaneously turn from mist-dripping quiescence to full blown riots as feeding flocks of Chaffinches, Bramblings, and Coal, Long-tailed, Blue and Great Tits, ricochet around our countryside.

By searching the leaf litter of any Oak wood you will come across the focus of their many attentions. The tiny flattened flying-saucer shaped growths of **spangle galls**, of which their are several different types, develop on the underside of leaves throughout the summer and now they drop to the woodland floor either detaching themselves or drifting down with the leaves. Once secreted among the leaf litter most will complete their development and emerge in the spring. However, though they appear hard on the outside, to many birds they have an irresistible soft and chewy centre.

Ground force

A crack squad of mean-looking armoured slug mashers are afoot at night. Cool wet nights mean invertebrates are on the move. Although it may not be warm, the moist autumn air is certainly more favourable than that of the dry summer and with the rehydration of our gardens, wasteland and woods, **ground beetles** are patrolling. Most – over 60% – are nocturnal, so

ABOVE: Spangle Galls
The galls look a bit like little flying saucers. They are the individual homes to tiny wasp grubs, which are a well known snack for birds. Flocks of Bramblings and Chaffinches feast on them over the winter months.

the surest way of seeing them is to get out with a torch and do a bit of foraging for yourself – in lawns, log piles, compost heaps.

They can generally be recognized (although you'll need a good field guide to be sure) by a clean-cut appearance, long legs, big eyes and a powerful set of mandibles. Most are the perfect assassin, using their mandibles like a pair of bolt cutters, taking out and slicing up any invertebrate they come across from slug, snail and woodlouse – some even specialize in caterpillars.

Did you know . . . ?

There are more than 350,000 species of beetle in the world! The Carabid Beetles, otherwise know as the ground beetles, make up a huge and very successful family with around 350 species in Britain. *Carabus intricatus* is the largest British species and it is very rare – it is found only on Dartmoor.

ABOVE: Acorns
From little Acorns, mighty Oaks will grow, that is assuming squirrels, Jays, deer, Nuthatches, weevils and gall wasps leave them alone!

Violet Ground Beetle
Carabus violaceus

Carabus granulatus

LEFT: Ground beetles
Under torchlight ground beetles look black, but under natural light some, like the Violet Ground Beetle, take on wonderful metallic greens, violets and reds.

Plants

The most obvious manifestation of this time of seasonal transition is that the leaves of **deciduous trees** drop off! (deciduous is derived from the Latin and means 'to fall down'). All those millions of leaves, which rustled in their virgin greenness through the spring, maturing to reach peak productivity in the summer were acting in a way like little light factories for the trees that bore them. Come autumn they are now cell by cell being decommissioned for the winter.

The biggest show on Earth

But the leaves don't simply leave the tree: a good job too because if they did all of us who have **Japanese Maples** would be left wondering why we bothered buying them in the first place! Before they hit the deck and become fungi fodder, leaves change colour and the best bit for all of us is that this palette of russets, fawns, yellows and rouges has absolutely no purpose as far as anyone can work out. Autumn colour is an accident, a by-product, an outward manifestation of the chemical re-shuffling that is going on.

The trigger for this vibrant final fling is not temperature or rainfall (although these do have some effect on the outcome) but day length. In response to shortening days the plant cells produce a hormone called abscisic acid, which starts the production of a corky layer of cells at the base where the leaf attaches to the twig. This gradually blocks the plumbing that leads to the leaf and eventually seals it off before a puff of wind sends it groundward.

RIGHT AND BELOW:
Autumn colours
Get out there and enjoy the free show. The leaves are turning from green to oranges, browns, yellows and reds; nature is doing its multi-coloured last performance. As a bonus, you can scuff through the piles of fallen leaves.

Simultaneously, the greens of chlorophyll are slowly breaking down. This decomposition goes on all year with the plant simply manufacturing replacement chlorophyll. Speeded up by cooling temperatures at this time of year the plant stops making chlorophyll and the greens seep away.

As the all-encompassing green fades out, other more stable pigments, accessory cogs in the light machine, are left behind. Outstaged by the chlorophyll all year the chemicals called carotenes start to shine, the yellows of xanthophylls and the reds and oranges of carotenoids.

Trees, Glorious Trees

Tullymore Forest Park, Bryansford nr Newcastle, Co Down (344329) – Set on the slopes of the Mourne Mountains, the Department of Agriculture Forest Service Forest Park provides stunning winter colours with a stunning mountain backdrop.

Forest Of Dean (SO628095) – This particular forest is massive, straddling the border between South Wales and Gloucestershire. Despite being squeezed from time to time by human pressure, it still is a beauty and retains the 'old' feel, with a rich mixture of old and new, broadleaves and coniferous. Of particular interest are Beechenhurst Lodge for colour and sculptures and the Arboretum near Speech House.

Westonbirt Arboretum, Gloucestershire (ST863903) – Get the timing of your visit right and your eyeballs are in for a full on bombardment of colour. This collection of 18,000 trees covering 240 ha is a mixture of native and exotic the result being probably the best display of tree shut down in the UK! One particular part worth focusing on is the Silk Wood Autumn Trail. Tel: 01666 880220.

Glen Creran, north of Oban, Argyll (NN031474) – This stunning mixed deciduous wood, on the right week in autumn, can be a panoply of zinging colour as the Oak, Birch, Rowan and Ash put on their last blaze of glory. Numerous walks intersperse the wood.

Cambus O'may, nr Balleter, Kincardineshire, Scotland (NJ436024) – This is a stunning valley even in the dullest month. Even though the forest here is predominantly Scots Pine. The dark sombre green sets off the sparkling bright yellows, golds and browns of the scattered Birch, Rowan and Willow and the smoky tints of Silver Fir. Car parks and walks make accessing this top spot rather easy from the A93 Aberdeen to Ballater road.

Epping Forest, Essex (TQ412981) – Less than 20 km from the centre of London, this ancient woodland, all 2,430 ha of it, stand and have been standing since Neolithic times. As with such stately woodlands the flora and fauna here is diverse, particularly good at this time of the year for fungi. The famous Beech pollards are probably the loudest. There is, however, a good selection of other deciduous trees from Hawthorn, Rowan, Ash, Oak, Birch and Wild Service Tree.

New Forest, Hampshire (SU261058) – A well-known autumn spectacle is found in the fabulous displays of bedazzling Oak, Beech, Birch and the yellows and reds of our native British maple – the Field Maple – and the rhythmic belching of the forest's Fallow Deer. This second spectacle is best experienced early in the morning and the best place to witness the autumnal palette is Bolderwood Ornamental Drive.

LEFT: Field Maple
Growing on most types of soil, the Field Maple can be found in hedgerows and woodlands.

RIGHT: Whitebeam
Another well-kept
secret. Keep an eye out
for different fruits and
you could discover a
few new species.

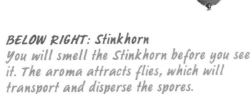

BELOW: Beefsteak
Fungus
Very common on Oak or
Sweet Chestnut, the
aptly-named
Beefsteak Fungus is one
of the most distinctive
of the fungal suite,
becoming visible in our
woods at the moment.

BELOW RIGHT: Stinkhorn
You will smell the Stinkhorn before you see
it. The aroma attracts flies, which will
transport and disperse the spores.

Waste products such as tannins, the brown colour of most **Oak** leaves, are also dumped in these earthward-bound waste crates. The crimsons that are so much part of the spectacle in the USA also get a showing here, although you may have to enjoy the spectacle in a single leaf or bush as opposed to a whole valley. Many non-native maples but also a few natives such as **Wild Cherry**, **Dogwood**, **Bramble** and **Guelder Rose** are all prone to a last minute blush. These chimney-fire reds and even purples are produced by yet another group of chemicals, the anthocyanins. These are created by the reaction of phenols (natural protection against leaf-chomping insects) with sugars produced by last minute blasts of photosynthesis but trapped in the leaf by the rapidly constricting abscission layer – a process favoured by dry sunny days and cool crisp nights.

Mouldering masses

In each wood there may be 200 million Oak leaves for every hectare. The bulk come cascading down (around 3 tonnes per hectare), with the first frosts and winds of autumn. They eventually end up back in the forest floor.

With the general dank, depressed moistness that is in the air at the moment it is not a bad

time to stop raking up and burning the crunchy mantle forming on your lawn. Abandon that broom you've been pointlessly chasing leaves around your yard with and go and find the hidden meaning of autumn.

Grab a good handful of leaf litter and turn it over, and under this year's crispy additions will be the flattened leaves of last autumn. Look closely and you will notice frail white strands, slightly fluffy in appearance, in places so dense they are like a lace-maker's sampler, in others just a single questing filament. Turn over logs, lift up pieces of bark and you will always find somewhere evidence of these mysterious threads, the weavings of **fungi**.

We tend to think of fungi being mushrooms and toadstools but the reality is for the most part invisible. They spend most of their life as threads of fungal mycelium. Like some unseen living telephone exchange, this bizarre pulsating network of cables is constantly moving, seeking, dissolving and shifting nutrients around the whole system, recycling the fallen glamorous stars of the woodland system. Each thread may be less than a millimetre in diameter, but they are so numerous that there may be 70,000 m of hyphae in a gram of soil, amounting to several tonnes of fungus per hectare. Like underground recycling Mafia, fungi run the joint.

It is in this damp season that the fungi mass together their hyphae into their fruiting bodies we call mushrooms and toadstools. It is the general dampness of autumn which allows this process to take place as fungi, whether mycelium or the fruiting bodies, have very thin cell walls and rely on absorbing large amounts of moisture to grow. Most of the year they skulk in the damp places but now with a superabundance of water they are not at such a risk of desiccation and send up a variety of spore dissipating structures.

With something like 6,000 species, it is a forager's paradise out there! Do not simply concentrate on whether or not you can eat them. Their numerous alien super structures are feast enough, inspiration for even the most jaded science fiction film set designer. The unreal red of **wax caps**, numerous **puff balls**, **Orange Peel Fungus**, **Jew's Ears**, **King Alfred's Cakes** ... just take a good field guide and go! But while you are identifying the glistening clump of eerie yellow that is a Honey Fungus, don't forget that what you are seeing is really the tip of a very big iceberg. Its hidden mycelia, which riddle the dead tree stump on which it sits, may occupy 15 ha, weigh 100 tonnes and be 1,500 years old!

ABOVE: Penny Bun
This is one of many very similar fungi belonging to the Boletus family, easily recognized by their plump look and a spongy sporing surface.

BELOW: Orange Peel Fungus
A non-edible cup fungus, this species grows among grasses or leaves on the ground. It can be found throughout Britain. A puff of spores looking like smoke are released as the asci open simultaneously.

November

Winter tightens its cold and clammy grip on wildlife this month. The naturalist may feel tempted to retire indoors to sit by a cosy fireside, but there are many spectacles that reveal themselves only now to those with the inclination to look. Birds clump together – Starlings, wagtails, tits, gulls and waders are all present in startling aggregations. The leaves are dropping off every deciduous bush and tree and with the removal of visual obstacles, nests of birds and mammals are revealed, as are their runways that, for the summer, remained hidden in the burgeoning clumps of grass.

ABOVE: Birdwatching
There's flocks of hundreds of birds to see, and it can get a little confusing to tell one little grey bird in the distance from another. Grab a telescope and do your best.

RIGHT: Nut cases
Protein-rich Hazelnuts are important food for a variety of creatures. You can tell what animals have been around by examining the empty shells.

Wood Mouse - marks of upper teeth evident.

Dormouse makes a smooth-edged hole.

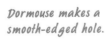

Squirrel splits the nut open.

Bank Vole leaves clearly delineated gnawed edge.

Mammals

Seeing big mammals is difficult at the best of times in Britain; seeing small ones is virtually impossible unless you put in a lot of effort. But it is reassuring to know that they are there and for the mammal detective it is fortunate that quite a few of them have a penchant for Hazelnuts. This year's fresh **nuts** are greeny white in colour while old ones of previous years are dark brown. By sticking your head into the half light at the base of a hedgerow, fresh feeding signs will stand out almost as of they are have a light source all of their own. Places to look include under root overhangs, under logs and objects such as corrugated metal sheets.

Separating the species then is a matter of looking at the way the nut has been opened and the tiny etchings and chips left on the surface by the consuming mammal.

Small mammals hold the nut against the ground and gnaw a hole, working the edge furthest away from the animal and turning, enlarging the entrance and chipping away at the kernel at the same time. Each has its own technique. **Woodmice** and **Yellow-necked Mice** create a hole with a chiselled edge created by their lower incisors, whilst those on the upper jaw grip the nut, causing a ring of scratches around the hole. **Dormice** do a similar trick but are finer engineers: they still make tooth scratches on the surface but leave a smooth almost chamfered edge to the hole. **Voles** modify the technique and stick their heads into the hole, nibbling at the near side. This leaves behind a rough-edged hole with no scratches on the outside at all.

Standing out

Brown Hares are seen by too few people. They are quite scarce animals but they are also well camouflaged and rather secretive. So what hope have you got of seeing its close relative the **Mountain Hare**?

This month nature's survival strategy for this animal is to turn them from their blue (actually browny grey) summer pelage to that of a complete wintry white out. This works to our advantage because a cruel twist of fate means that snows may not have fallen yet and so in many of their upland habitats the Mountain Hares stand out like, well like white hares on a dark background!

At this time of the year most have completed their third and most spectacular moult of the year: from their dark summer coat they change to white or skewbald with black ear tips. The transformation is probably triggered by day length. Different individuals moult at different rates, starting at the flanks and working forward a white wash of their longer more insulating winter coat. There is a lot of variation – some lowland populations do not turn at all, while even in upland areas the transformation is often not complete.

The best places to see Mountain Hares are the Scottish Highlands and the Peak District, where there is an introduced population. These hares undergo the full white treatment and a walk in the hills can often reward you with sightings of these animals 'glowing'. They stand out so much you can often spot them from half a mile away and in places in the Highlands they can be so numerous that to see one hundred on a single hillside is not unheard of.

LEFT: Squirrel nutkin?
Squirrels have scale on their side. Using their substantial blades they chip a hole in the apex of a Hazelnut before using their lower teeth to slice the nut in two.

What to look out for this month

- Signs of gnawing
- Mountain Hares
- Starling roosts
- Salmon spawning
- Pupae in the earth
- Ivy in flower
- Ferns and spleenworts

BELOW: Mountain Hare
Mountain Hares are also known as Blue Hares, after the grey summer coat in some populations. In Ireland they are the only hare, so they are known as Irish Hares. It is smaller than the Brown Hare and has shorter ears. This one is mid-moult, showing its white coat and the remains of its grey one.

Footprints

Wood Mouse

Brown Rat

Grey Squirrel

Muntjac Deer

ABOVE: Footprints
A selection of mammals' footprints, all roughly to scale.

RIGHT: Deer tracks
Deer tracks are known as 'slots'. It takes a bit of practice to separate the deer species from each other and from Sheep, the bane of the deer tracker's life.

ABOVE: Stoat
In good company with the Mountain Hare and Ptarmigan, Stoats in the north of the British Isles moult into an ermine winter coat.

Standing on a Scottish moor, shivering, I was watching a rock-strewn hillside when the rocks got up and moved, but not slowly as you would expect from a rock but as if an invisible force was hurling them, they tore off up the slope and vanished over the horizon. On scaling the escarpment and peering into the next valley I found myself looking at a large number of hares.

Other creatures that turn white are **Ptarmigan**, a species of **upland grouse**, and northern races of the **Stoat**.

Prints of the night

Just about every animal that walks, creeps and crawls blows its cover in some way, however small the evidence. An observant scurry through park, woodland or field should reward the Holmeses and Marples among us. **Footprints** are the obvious clues that spring to mind, but are often hard to find and interpret. Life can be made easier by concentrating your efforts on areas of soft soil. Rivers and ponds have mini-beaches that can provide rich insights into the movements of the most cunning and secretive of creatures. Fine sand and silt banks on a riverbank can be peppered with spidery star-shaped prints of **Mink**, rat, **squirrel** and **heron**.

Practical bit

Muddy magic

Now is a time to start a footprint collection. For big prints of mammals and birds look along footpaths, edges of puddles and streams. Keeping them is easy. Look for good specimens in firm mud or sand, surround with a hoop of cardboard held together by a paper clip or staple, then mix up in a pot a nice runny mix of plaster of Paris. Make sure it is runny enough to drip off the mixing spoon, but not so wet it takes hours to set. Pour it gently into the mould, leave to set for 15 minutes before lifting the whole lot up and cleaning off any soil or leaves.

An excellent way of creating your own portable substrate is by coating a sheet of glass with carbon. Hold a piece of glass with gloves about 2 cm above a candle flame. Place this carbon-side up anywhere you fancy overnight – garden sheds and log piles are often good locations. By the morning you should have some scratchy prints of small mammals. This method is quite sensitive, so you may even pick up the tank-like tracks of a centipede, scuffs of a spider or the pockmarks of a beetle. If after the first few nights you have nothing, try baiting with seeds for micro-veggies such as mice and voles, or a dab of cat food for shrews.

Birds

In the summer there are 'Little Brown Jobs', known as LBJ's. More often than not these are warblers. Now it is winter there is another reason for the birdwatcher to hide between the pages of his field guide. It's time for the return of the Little Grey Jobs! More lovingly known as **waders** in winter plumage, they are not simply grey but a subtle black, white and buff filigree of feathers. It's just that as soon as they start tearing around in the sky in flocks of hundreds, like some ornithological tempest, or shuffling and re-shuffling in the wind or on the water's edge, it can get a little confusing to tell them apart to say the least.

So what if you don't feel confident about separating your **sandpipers** from your **Sanderling**? You just simply have to bite the bullet. You can stare at a field guide all you like until you know all the identification features off by heart, but applying it to reality is very different. The best way to get a handle on the little grey jobs is to spend some time in a hide looking at them. I say hide because I'm a bit of a wimp, any excuse to get out of a biting November wind, I'll take it. Hides can be very comfortable places and if you are going to unravel the numerous wading birds in their winter plumage then you may be some time so being warm and dry means you can concentrate on the grey job in hand rather than worry about

whether or not your binoculars are waterproof. Hides are also excellent places to meet other birdwatchers, some of whom may be having the same I.D. problems as you or even have the answers, and five minutes talking may be worth, in information, a million field guides to you.

Ideal home exhibition

This month all but the most stubborn of Beech and Oak leaves have now departed from their summer anchorages. The trees not only reveal their own specific profiles but among their branches are betrayed many of the summer's secrets. Birds' **nests** are revealed; it is taboo to go looking for them in spring and high summer when they are fulfilling their function but now they lie vacant of birds, and it is a good time to revel in their form and function. Dense bundles of twigs and vegetation give away where birds' nests were built so if you failed to find out where that **Long-tailed Tit** was heading with all those strands of horsehair and moss way back in the summer, now is your chance to become a retro-twitcher!

At first one nest may look very much like the next, just a tangle of sticks, straw and mud, but with a little experience, a touch of guesswork and the usual healthy helping of dogged persistence,

ABOVE: Long-tailed Tit's nest
It is usually taboo to approach birds' nests at any other season, but now it is fair to assume that they have been deserted, although some of the domed affairs may be used by small mammals or birds as a roost site.

LEFT: Knot and Dunlin flock
If you have never gone estuary birdwatching in the winter, then you really must experience the bustle of these places. There are plenty of locations to choose from in the UK. With over 14,500 km of coastline, there are over 160 intertidal top spots scattered all around our coasts.

ABOVE: Rookery
Rookeries stand out even at a distance, but unlike the nests of many other birds these still have a function as a base for the colony of Rooks, which will be doing some renovation over the next months in preparation for the coming breeding season.

you can soon start connecting these constructions to their ornithological originators.

Distinctive nests such as the large loose-domed stick nests of **Magpies**, the tree top communities of nest platforms constructed by **Rooks**, and the solitary efforts of **Carrion Crows**, are relatively easily seen even from a distance. A slightly more taxing challenge are the similar sized nests of various commoner garden birds such as **Blackbird** and **Song Thrush**, which can be separated on constructional merits. Song Thrushes are unique among British birds in having a hard lining of mud to their nests, while Blackbirds use mud in the construction but line the nest with fine grasses.

Close to stumps and in dense vegetation like moss footballs with a hole punched in the side are the masterpieces of **Wrens**.

These are the easy ones. **Finches** make nests of finer materials than the thrushes – small cups of grass, hair, wool and moss. The **Greenfinch** and **Chaffinch** are less fussy than the others as to where they site their residences and their nests are most likely to be found in a garden hedge. For the other finches you need to look a little higher – in the forks of trees close to the trunk or towards the ends of branches.

Look out for stashes of seeds and fruits, as hedgerow nests are often used by squatters such as voles and Woodmice.

RIGHT AND FAR RIGHT: Woodpecker nest holes
Different species can be pinpointed by the size and location of their holes. The hole in the Birch tree is typical of Green Woodpecker, while those of Lesser Spotted tend to be in cavities in a soft trunk or on a thick branch.

Keep a watch out for the nest holes of woodpeckers. **Green Woodpeckers** have the largest nest entrance of around 7-8 cm in diameter, the **Great Spotted** being of similar dimensions and the **Lesser Spotted** a tiny doorway of about 4 cm, but there is more to this than simply the size of the holes! Green prefer to knock holes in healthy looking trees with rotten hearts, Great Spotted tend to use trees that are obviously on the way out and Lesser Spotted are often higher and on the underside of a sloping branch.

However, it is not just the architectural activities of our avian fauna that you are likely to come across whilst scanning the lofty levels of a woodland. You may notice the summer drey of a Grey Squirrel, but the chances are that you wouldn't know you were looking at one as they resemble a hollowed out crow's nest built high and out on the branches. Much more distinctive is the dense winter drey that is also used as a nursery. These are often much larger, constructed with leafy twigs, lined with mosses and grass and built close to the trunk where they are less vulnerable to buffeting winter gales.

Pies in the skies

Not immediately suspicious of the fact that there was actually a space in the car park, conveniently close to the station doors, I parked up. Lady luck was smiling on me... I thought.

Lady luck wasn't the one scratching off the hundreds of white spots of bird excrement the next day. My car had been reduced to a monochromatic Jackson Pollock by several hundred innocent looking little bundles of feathers – I had found a **Pied Wagtail roost**.

With a taste for urban life, Pied Wagtails are one of the most accessible of the many birds that form roosts during the winter. Numbers swollen by Continental birds, they too spend their days feeding in the country, playing fields and parks, before coming together in pre-roost flocks at dusk to descend in a single 'Hitchcockian' movement into a street tree or a selected rooftop.

Just about every urban centre has a roost, so to find yours simply stare at the twilight sky, plotting the movements of pre-roost flocks. It may take a few days of chasing and mapwork but with any luck you will find your pile of 'Pies' in the end.

Some birds collect together to exploit concentrations of resources. Where there was berry and haw yesterday, today there is just a black mesh of thicket and thorn. Pillaging and plundering flocks of **Redwings** and **Fieldfares**, mixed

LEFT: Pied Wagtail
Pied Wagtails congregate in spectacular numbers in various urban locations.

BELOW: Fieldfare flock
The Vikings are invading again! Fieldfares are on the move back into Britain from Scandinavia. Listen for their noisy, clacking calls and look out for their large size and distinctive grey rump.

RIGHT: Redwings
My favourite thrush, the Redwing, often joins other thrushes at the feast of tree berries. Listen for their 'seep seep' calls as they move over late at night. Like Fieldfares, they are mostly winter visitors to Britain from Scandinavia.

OPPOSITE: Starling flock in Brighton
Starlings have always been truly underrated. Pretty, intelligent and bold in the singular, flippin' awesome in the plural, especially in their huge winter roosting flocks like this one at Brighton's West Pier.

with a peppering of the more familiar resident thrushes strip fruits from our hedgerows.

Bundles of birds

There are many ways to increase your chances of seeing some of the great birding spectaculars. Many wintering bird species form massive eyeball-busting flocks.

About now an evening visit to your nearest **Starling roost** becomes a must. These often maligned but very successful birds, usually hardly noticed if we see them waddling gawkily across our lawns in ones and twos, have their numbers swollen this month by birds hopping over from the Continent. It is a bit of a mystery why they form such large flocks: possibilities include safety in numbers, increasing their foraging efficiency or simply for warmth. The noise and sight of seething masses of thousands of birds swirling like a shoal of herring in the sky make this among the most spine tingling of scenes.

Feathered fantastic

The din of thousands of squeaking voices, the whirr of wings tearing at the air and the giddy rotation of the perpetually swirling smoke of birds, leaves me feeling elated if a little motion sick. This is how it felt to witness a Starling roost at Brighton's West Pier one crisp evening.

It is only now in the British winter that you can really appreciate some of the most frequently witnessed and indeed the most awesome wildlife experiences – a spectacle that is the flocking of birds.

Did you know . . . ?

Grub's up!
The way to a **Robin's** trust is via its stomach and even though the thought of a handful of wriggling Meal Worms may not seem appetizing, Robins find them irresistible.

Practical bit

Teasel the birds

Try something different in the garden. Rather than putting nuts in feeders and crumbs on the lawn, attract a few more finches by collecting Teasel seedheads, and sprinkling tiny Niger seeds into the natural seed chambers – this turns them into recyclable seed heads, whilst in the wild they would only be used once! You could experiment by melting fat into some as well – **Blue Tits** particularly like feeding on this.

In order to attract the maximum number of bird species to your garden, it is important to make food available in a variety of forms and locations to suit different species' feeding strategies. Many birds prefer feeding on the ground or taking food scattered freely on a flat surface. Reserve an area of lawn or border for this style of feeding.

Hanging up baskets of peanuts, sunflower seeds and the like will also attract a nice wide range of bird visitors to your garden.

LEFT: Goldfinches on Teasel
Goldfinches are a fickle bunch to persuade to visit your garden, but they simply cannot resist Teasels. Once empty, refill the seedheads by pouring niger seeds into them.

137

Hot Spots for Starlings

Below is by no means a definitive guide to all the Starling roosts and most towns have a roost of some kind. Other good ones include: Manchester Piccadilly Square, Brighton Pier, Bradford City Centre, Runcorn Bridge.

Leighton Moss, Lancashire (SD469752) – This is quite simply a cracking spot at any time of the day at any time of the year! In November it is stuffed to the gunnels with winter duck, but what better way to end a day winter 'wildfowling' than to watch the Starlings descend on the reedbeds to roost, often escorted by a shady chaperone, a hunting Sparrowhawk?

Aberystwyth Pier (SN581819) – This particular roost of up to 10,000 birds exploits the many ledges, nooks and crannies provided by this local landmark. Watching this roost is best done while eating local fish and chips from a newspaper.

Slapton Ley NNR, Devon (SX826431) – A Starling roost has been recorded here for some 200 years. A visit 30 minutes before sunset from the end of November is almost certain to reward you with a large tempest of Starlings as they fill your senses before settling down in the reedbeds.

Glasgow City Centre (NS587651) – When you've finished your days shopping in Argyle Street, it is worth lingering until twilight in George Square with coffee in hand. Here you can witness many thousands of Starlings which gather in boisterous moots on and around the ornate Georgian facades.

Forth Bridge, Edinburgh (NT124807) – This is a classic urban roost – not as large as it used to be like many around the country – but the natural spectacle combined with the manmade splendour of the bridge itself is stunning. If you catch all this in front of the fire of a winter sunset, the memory will never leave you.

Huddersfield Station (SE143168) – This is yet another classic urban roost that still manages to 'cut the mustard' as far as sheer numbers go. Not as large as 10-15 years ago but still one of many thousands and well worth missing your train for.

Poole Harbour, Dorset (SZ032877) – This is a large area to look for the birds over, but the most likely spot is around the mouth of the harbour. It can be quite a good bit of detective work to follow the smaller congregations or 'moots' over a couple of days and from their direction of flight eventually trace the main roost.

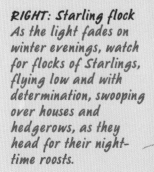

*RIGHT: Starling flock
As the light fades on winter evenings, watch for flocks of Starlings, flying low and with determination, swooping over houses and hedgerows, as they head for their night-time roosts.*

Amphibians, Reptiles and Fish

A fish out of water is like a wet pebble drying. As soon as a fish is torn from its little-understood world, either by net or by hook, it instantly loses the lustre of the live animal in its element.

Fresh salmon

Fish-watching is fraught with all sorts of problems. For a start they live in water, an often murky element. Even if it is clear, any views are often hampered with surface reflections, weed and the often skittish nature of these nervous animals – a quick tail flick and it is gone.

However, taking place in British rivers is a spectacle that is about as dramatic as fish behaviour gets anywhere in the world and what's more, it's relatively easy to observe – the **Atlantic Salmon** is spawning.

With the first heavy rains, many of our rivers rise in spate. In estuary mouths the salmon, which have been congregating by some rivers for many months, leave the saline world that has been their home for the last one to five years and embark on one of those wildlife epics, a massive hormonally fuelled journey upstream towards the gravelly, oxygen-rich headwaters of their natal river.

Leaping for life

The autumn freshets or spates lift the water levels high enough to carry the Salmon over obstacles. Rises in river levels may be short-lived so progress upstream may be sporadic. When a spate is on the best Salmon-watching is undoubtedly to be had at weirs and waterfalls.

The fish hurl themselves up walls of water a couple of metres high or more at a rate of a couple a minute. On my local river this has become quite a spectacle, often drawing crowds of people from those with pushchairs to pensioners, all shouting encouragement to the fish!

BELOW: Salmon on the run
During this upstream surge the Salmon change, becoming super fish, some attaining a full metre in length.

RIGHT: Leaping Salmon
Atlantic Salmon are the only fish I know of in this country that can attract a crowd of people to watch. Check out a 'swim past' in their incredible odyssey. Weirs, dams and waterfalls are some of the best places to watch them muscle against the current, putting into perspective their achievement. The males change their look - they become less silvery and put on their battle dress, their flanks blush a reddy-pink and their heads change to a fearsome battle club.

Change of face

On quieter days the Salmon can be watched (best with Polaroid sunglasses) stacking up in the calmer, deeper waters, either resting, or sometimes waiting for the cover of darkness to move (a behaviour associated with clearer watered rivers). They do not feed during their sojourn in fresh water, relying on reserves laid down in the ocean.

On the way up to the clear headwaters they undergo a metamorphosis that is most pronounced in the males. As well as losing the silver of sea fish they put on battle dress, their flanks blush with all the colours of tartan and their heads change from the bullet-profiled utility head to a fearsome battle club; the jaw bones elongate to form a hooked kype and the teeth grow longer.

Whilst these dramatic changes take place externally some pretty major ones occur internally at the same time. The reproductive organs develop, the gut shrinks and becomes useless; these migrating fish fast for the journey, which could in some instances take up to several months to complete.

Reddy to spawn

For me the cream of Salmon-watching occurs right at the top of the river. Here the water is clear and crisp and constantly dancing, even when the land above the surface is locked in frost, with nothing but the occasional pipit as evidence of life. There it is possible to witness the culmination of this journey which could have taken the fish thousands of miles.

The spawning areas can be action-packed. In the shallow waters, hens (as the females are called) are chaperoned by one or more cock fish. Their slow, deliberate shadowing of the females contrasts with the energetic barging and thrashing of competition as they rally for the best spot and the best females in a few inches of water. Breaking the surface, fish this size simply look incongruous in such shallow water, often running themselves aground in the process.

The hen carves the redd or spawning furrow in the gravel of the river bed. She lies on her side and thrashes her flanks against the gravel, forcing pressure waves in the water that waft the gravel aside. If you read the books they make it sound as if making a redd is a small project, after all how much landscaping can a fish do? In reality it can be as deep and as wide as a wash basin. The eggs are deposited here (approximately 900 per 500 grams of fish weight). The cock sheds the milt (sperm) over the eggs. The pair repeats the process four or five times, slowly moving upstream as they go.

Invertebrates

There is nothing more pleasing than a moth **pupa**. All shiny and red, brown or orange, many spend the winter months in this stage of their lifecycle between caterpillar and adult moth.

Go out and try and find one of these hidden away underground. Take an old margarine tub of soil or moss, a trowel and a hand fork and gently dig around the bases of trees. Do not tear up the soil or go any deeper than 10 cm.

Be patient and you will find the 'gems' you are after. In my experience certain trees are better than others. The best are the willows, Sallow and Poplar (you have to become a bit of a tree expert too!). It is here that you stand a chance of finding the big one, the shiny bullet-shaped sarcophagus of a hawkmoth – such as a **Poplar** or **Eyed Hawkmoth**.

Keep an eye out for the **Comma** butterfly. These will still be flying wherever there are flowers; Ivy is still flowering (see page 142). Comma butterflies will settle on Ivy to feed. Once deep, cold winter days set in, then this insect will seek out shelter in a hedge or tree to hibernate.

In order to survive the harshness of winter, it is usual for a sheltered spot to be occupied by hibernating insects. Favoured places include thick, tussocky grass, inside the dead stems of certain plants, beneath the bark of trees. Most species of the ladybird family seek a sheltered spot, clustering together in a group.

RIGHT: Hibernating Ladybirds
Ladybirds sit it out in crevices under bark, in garden sheds and cool window sills.

ABOVE: Moth pupa
The creepy-crawlies are down but not out. Just pick around in the leaf litter for the pupae of moths.

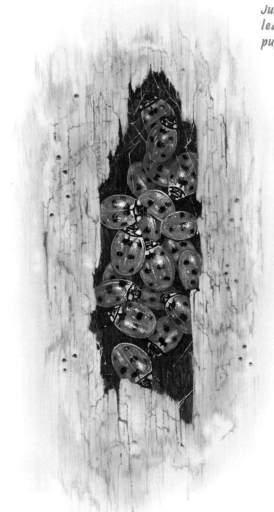

Practical bit

Hatching pupae

If you are curious as to exactly what you have got, sterilize some soil or moss (10 minutes in a microwave oven), place in a seed tray or tub and arrange your pupae collection on this. Keep in a cool place such as a shed or garage out of the sun and spray with water every week, making sure that the soil is always damp but not soggy. Place a few sticks in the soil so the moths have something to climb up and watch and wait. Most will probably hatch in the spring, but some, such as various winter flying moths, may surprise you, so check them every day.

Plants

RIGHT: Red Admiral on Ivy
Yes, believe it or not, assuming we haven't had too many cold snaps, there can still be thirsty insects on the wing. Late migrants like this Red Admiral butterfly make the most of late-flowering Ivy.

Just when we think that the buzz of gauzy wings is a thing of past months, November can surprise us with a gentle warming of weak sunlight, enough to fire up the thermal slaves still after a quick fix. Suppliers of that insect elixir of life, nectar, are in short supply, but one flower still juicing up is **Ivy** and this and the combination of its habit of draping itself over walls and tree trunks means that it often finds itself in sheltered sun spots that concentrate the little heat still left in the sun.

Those Ivy stems bearing flowers have a more rounded appearance than the classic lobed leaf of the clambering form, while the flowers themselves are arranged on little candelabra. What the yellow-green flowers lack in flamboyance they more than compensate for in decadence, producing large quantities of nectar, which attracts any late insects.

On sunny days they can bring in the usual buzzing throng of flies, a late royal visit from a queen hornet or common wasp, Honey Bees, and possibly the dusty wings of Small Tortoiseshell butterflies.

One unseen resident of Ivy this month is also the Holly Blue butterfly, even though the caterpillar itself should have fed up on the flower buds a month or so ago. The chrysalis containing the ingredients for next spring's generation of butterflies is sitting out the winter on Ivy leaves and twigs.

Rock gardens

Anything that clings to its chlorophyll this month becomes instantly noticeable. But turn from macro thoughts to micro, and there is a world of intrinsic beauty that will outshine even the most elegant of evergreen trees. The **wall ferns** and **spleenworts** are simple and elegant little plants that have lain shrivelled and sun-baked in deserts of rocks, walls and mason-

Practical bit

RIGHT: Parasol Mushroom
The Parasol Mushroom is a beauty. Check out the scales, like wooden roof tiles, and a loose collar around the stem.

Spore print

Fungi do not have seeds; they have tiny smooth, wind-borne spores. Every mushroom or toadstool has millions of them. Collect an umbrella-shaped fungus; cut off the stem close to the top and, in a draught-free room, place the fungus, gill-side down, onto cardboard or paper. Cover with a bowl (to stop draughts blowing the spores) and leave overnight. Lift the fungus up to reveal the shapes of the gills made by the spores. To keep your spore print, spray hairspray onto your print from 50 cm away and leave to dry. Repeat 3-5 times.

Make a collection of different kinds and use dark-coloured paper for those with white gills and light paper for those with dark gills. Don't forget - some fungi are poisonous so wash your hands well afterwards.

ry for the summer months. They have recently been rehydrated and rejuvenated by the moisture that comes with autumn. Now they bristle out from their cracks and crevices, turgid, pinnate and lovely.

Well, I say pinnate, but one of the most easily identifiable is the only species that isn't. The high gloss, patent green of **Hart's Tongue Fern** *Asplenium scolopendrium* is also a species that seems to look good throughout the seasons. This could be due to its very waxy leaves, which mean water loss is minimized and could also be reflected in its distribution. Tolerant of a variety of habitats, it is one of the most common and widespread of the *Aspleniaceae* family.

The other, more Lilliputian, members of this family are true wall dwellers, and have taken to crumbling masonry and dry stone walls in addition to their natural haunts of outcrops, scree and cliff. In the south and west the alternately lobed **Rusty-back Fern** *Asplenium ceterach* is unmistakable if you flip it over, as it has an almost felt-like covering of scales on its buds and on the underside of the fronds.

Did you know . . . ?

The Rusty-back Fern *Asplenium ceterach* is possibly the reason for the spleenwort family's common name. 'Wort' has its root in the old English word for herb 'wyrt' and 'spleen' comes from the use of this species to remedy spleen and liver disorders.

Other common species are **Wall Rue** *Asplenium ruta-muraria*, the **Maidenhair Spleenwort** *A. trichomanes*, a delicate species that has a resemblance to the Maidenhair Fern, commonly grown in bathrooms the world over. Its black main stem makes it easy to separate from the **Green Spleenwort** *A. viride*, which is similar but with a green mid-rib.

Fern green fire
To match the big colour show put on by our deciduous trees every autumn, the tiny un-fern-like *Azolla filiculoides* changes from its lush velvet green to a stunning red mass, turning the surfaces of ponds to fire, slightly making up for this alien's tendency to choke ponds.

December

If you suffer from seasonal depression the popular view of December is one that will keep you walled up in your protective, centrally-heated envelope for at least three months. Wake up! There is a lot more to this month than Robins and Mistletoe. All you have to do is prepare for the worst meteorological onslaught you could imagine, don boots, fleece and waterproofs and get out there – you will be surprised!

Mammals

The frustrating thing about British mammals is that the majority of them are small and most of them are very shy with a tendency to scuttling – habits that are not conducive to obtaining good views.

Consequently the best sightings we often get of them are either fleeting glimpses of furry, tailed backsides or completely unexpected meetings that leave both parties as surprised as the other. For the mammal watcher, winter can

ABOVE: The seven sleeper
The Common Dormouse gets its head down for a winter of hibernation. Do not be tempted to do the same - you will miss out on a lot of fine wildlife.

LEFT: Mammal table
Make your own mammal table from a piece of board raised off the ground. Who knows what furry visitors you may attract?

make things a little easier, snatching away the blanket of ground cover which conceals most activity throughout the other seasons as well as putting the squeeze on furry stomachs, meaning the creatures have to spend more time foraging for food. Even though this increases the odds of stumbling across a hunting stoat or shrew, seeing mammals can still be a bit of a hit-and-miss affair.

Patio desperadoes

By creating the mammal equivalent of a bird table (see below) you can turn the odds back in your favour, allowing a little more than a quick glimpse of your own backyard's furry fauna.

The sorts of mammals you are likely to attract, such as **Field Voles**, **Bank Voles**, **Wood Mice** and even **Yellow-necked Mice** are all of a particularly nervous disposition, a trait that is understandable when every Fox, cat, and owl in the neighbourhood is on their case.

It is worth experimenting with types of baits: seeds, nuts and fruit such as apples tend to be favourites. This always works best if you pre-bait the area on and around your table for as long as possible before you sit down to your first watch as by then most of your local inhabitants should have cottoned on to the free food.

Wait after dusk, wrapped in your 15 tog thermals and armed with a good torch with a red filter of cellophane. If it is a clear night, after about 20 minutes – about enough time for your fingers to freeze to your torch and the blood to drain from all your extremities – you will probably be needing something to distract your mind from the discomfort.

Two nighttime classics are on the winter air waves about now, normally reserved for TV atmospherics: they can now be experienced first hand. Tawny Owls are pairing up and establishing territories, so these birds are probably at their most vocal and, sitting in the dark, you can listen in on the throaty flutings produced by both sexes and the 'kewick' call of the females. The trained ear will be able recognize individuals, all of which have certain subtleties to the basic pattern of the song. The other distinctive sound belongs to foxes.

An often surprising sound mixed in with these stalwarts is the 'seep seeeping' of migrating flocks of Redwings as they pass overhead.

LEFT: Yellow-necked Mouse
The biggest of the bunch! If this large and rather arboreal mouse invades your loft you may assume it is more closely related to kangaroos than to mice!

What to look out for this month

- Foxes
- Mice and voles
- Buntings
- Wildfowl
- Hibernating insects
- Lichens
- Fungi

BELOW: Bank Vole
Round features, small ears and short tail make it easy to tell voles from mice.

Practical bit

Mammal table

To reassure nervous mammals like voles and mice, choose a location well. Select somewhere close to natural cover such as a hedge and position a piece of board, raised slightly off the ground to make viewing easier and to protect small visitors from those ever-present predators. If you cover it with a ball of chicken wire, you will keep the bait together. The aim is to prevent the particularly highly strung animals and those with hoarding tendencies from perfecting a 'grab and run' technique.

ABOVE: Amorous Red Foxes
It is common to see two foxes together in December - dogfoxes are desperately seeking vixens to mate.

Foxes flaunt it

Now is the season that **Red Foxes** will make their presence known. December and January are when they become more visible, more audible and much more odiferous!

The reason every dogfox is particularly focused is that the vixen of his dreams will be receptive for 2-3 days only. This is why it is common to see two foxes together now; the vixen being tagged by her dog.

So that every fox knows the position and reproductive state of everyone else, they communicate in a number of ways. As they go about their manoeuvres foxes urinate at regular spots. They also deposit their droppings in prominent places such as tree stumps. If you get close to the fresh path of a fox you may catch a whiff of a flowery smell – this is from the glands on the fox's brush – the violet glands.

The other way they stay in touch is by voice. I remember as a child, waiting awake in my bed waiting for Santa to deliver his booty. It was just when my eyelids were winning the battle with my brain, that they were flung open again by an unearthly wail. Then there it was again right outside my window. All my thoughts about the jolly chap in scarlet were replaced by another vision I knew to be red; the spirit-like form of a fox drifted across the lawn and away over the fields, calling every few minutes.

It is not only the almost human scream of the vixen (female) that you should listen out for, but the deep very dog-like trisyllabic 'bow, wow, wow', the contact call of both the dogfox and the vixen and a whole range of mechanical clicks, murmuring and 'yips'.

You may also be lucky enough to see a vixen leading the dogfox on the pre-nuptial follow-the-leader. Mating occurs soon after this event and if you count 53 days on from witnessing this scene you can predict roughly when the cubs will be born.

RIGHT: Fox screaming
Foxes are getting vociferous as the breeding season comes into full swing. The 'wow-wow-ing' bark and the blood-curdling scream, produced mainly by the vixen, are their way of getting the lowdown on who's who and where.

Birds

With their populations diluted and spread out for the breeding season, this family of birds is often overlooked. However, at this time of the year, like many small birds they go in for flocking.

The family I refer to are the **buntings**, of which we have six species that can be seen during the winter months. Their body shape resembles a thickset sparrow; this is reflected in the 17th Century word 'bunting', which means plump or stocky.

The **Yellowhammer** is the boldest of the bunch, the one that everyone who treads in a gorsey or grassy place during the summer months will know from its 'Little-bit-of-bread and no cheese' song. In the winter it does little of this and is also not quite the lemon-yellow 'canary'. But it does form large flocks to roost and if you stumble upon a group of birds that might fit the description have a closer look as among the yellower birds with the russet rumps, you could well find the much larger **Corn Bunting** (an unspectacular brown on brown streaky affair). The **Reed Bunting**, not in breeding plumage but sporting a characteristic white 'eye-brow' and moustache, is fairly widespread and a more likely roosting companion of a posse of Yellowhammers.

All of the above can be encountered on pretty much any habitat that will give up seed – rough grassland through to arable set-aside. But for the rest you need to look elsewhere. **Snow Buntings**, regular in and around the bins high up near the Cairngorm ski-slopes, can be found wandering the strandline pretty much anywhere on the coast. When seen in a big flock their pale washed-out plumage makes them look like a flurry of winter snow.

This leaves us with the most exclusive two of the tribe. If you cannot get down to the South Devon coast you will not see **Cirl Buntings**.

RIGHT: Snow Buntings
A winter flock of Snow Buntings will warm you, even on the coldest of days. White and black flashes from winter males really do look like a flurry of snow.

This is the only place they are found and quite frankly if you intend to make a special trip leave it until the spring when the birds are singing and the males have shed their dull streaky winter attire in favour of a black beard and eye shadow on a custard yellow foundation. The sixth species, the **Lapland Bunting**, is so rare and of such uncertain status it's best left for the bunting boffins.

ABOVE: Winter flock of Yellowhammers
The need to find food often concentrates small birds into highly noticeable, mobile flocks. This can make them a lot easier to see than in the summer.

female *male*

RIGHT: (From back to front:) Blackbird, Redwing and Fieldfare
Winter thrushes often flock in large numbers and their noisy 'clacking' makes them one of the most accessible seasonal spectacles.

ABOVE: Treecreeper
Treecreepers find that the soft, spongy bark of the introduced Wellingtonia tree makes a snug bed. Look for white splashes of droppings to reveal a roost site when the bird is not at home.

Storm cocks and wind thrushes

These are names which sum up well the meteorological situation. Large clattering flocks overhead signal the peak of the annual Viking invasion of **Fieldfares** and **Redwings**. Once the cold weather sets in they pillage en masse just about any bush with berries such as Cotoneaster, Rowan and Holly, with a particular penchant for Hawthorn. All of these are commonly found in gardens and parks all around the country and this is a good time to practise your thrush identification as it is possible to get every British species except for one feeding together on a single bush. They can be quite nervous birds, but you can try luring them down to your lawn by leaving out apples, but even here they tend only to visit those farthest from the house. As always, the best views require experiencing a bit of pain and discomfort. Secrete yourself deep within a well-endowed Hawthorn bush that the birds have been visiting, and by sticking to those standard ethics of field natural history, wearing dull clothes and keeping very still, (easier said than done when sitting on thorns) you should be amply compensated.

Urban extra!

Another urban ornithological extra worth looking for requires first identifying a tree. Old parks and churchyards are the best places to find mature Wellingtonia trees; these unmistakable giant coniferous trees native to California have a soft deep and fibrous bark. Check this over in daylight and the chances are you will find small oval depressions in the bark. These are made more obvious by a trickle of white bird droppings below each one. Return on a cold night and you will find these plugged with the tiny tawny streaked bodies of **Treecreepers**. The birds hollow out these customized and insulated snugs themselves. A single tree can attract birds from all over the neighbourhood, seeking sanctuary from the cold, and as many as 25 can be seen on one large tree.

Wildfowl Wonders

The Mere, Ellesmere, Shropshire (SJ403348) – Shropshire's Lake District is a collection of glacial lakes left behind by the retreat of the last ice age. And in the series of nine lakes that comprise this site Smew occasionally turn up, but the presence of many other winter duck mean that a Smewless day is far from a lost cause.

Ballyronan Marina, Lough Neagh, Co Tyrone (949855) – Lough Neagh is the largest freshwater lake in Britain and Ireland and is one of the most important bird habitats in Western Europe. There are more Pochard, Scaup and Goldeneye on Lough Neagh in mid winter than on all other lakes in Britain and Ireland put together. This part of the Lough is probably the best place to see inland scaup in Ireland. Tufted Duck, Great Crested Grebe and Mute Swan are just some of the many wildflowl species seen here. Contact Ulster Wildlife Trust for a bird guide.

Chew Valley Lakes, nr Bristol (ST574614) – This large expanse of water is one that is easily accessed and has many hides which are more useful to provide shelter against the chilly winds than to hide you from the birds. In winter some large flotillas of the Goosander can be seen here mixing it up with many other winter duck, Goldeneye, Wigeon, Pochard and Teal amongst their numbers.

Fairburn Ings, Fairburn, West Yorkshire (SE450275) – This is about as far north as you are likely to get Smew as this bird has a south-easterly distribution in the UK. Most of these birds breed in Russia and winter in the Netherlands. Unlike Smew, Goosander can nearly always be seen here among a good variety of other duck.

Grafham Water, Cambridgeshire (TL143672) – Visiting this one on the chance of seeing Smew is a bit of a long shot but choose a bitter morning after a bout of easterly winds and you may well be in with a chance. A trip here is never wasted, with plenty of other winter duck and waders easily observable especially from the hide near the Mander Park Nature Trails.

Abberton Reservoir, Colchester, Essex (TL961185) – This is probably the nearest you get to a 'dead cert' in the wonderfully beguiling world of the Smew. But even here you need to be lucky, with your chances increased considerably with a cold Arctic snap and east wind. However, the large areas of water act as magnets for duck with Goosander often present as well as thousands of Wigeon and Tufted Duck.

Staines Reservoir, Surrey (TQ050730) – The elite in urban duck watching sites, the sheer spectacle of seeing 4,000-5,000 Goldeneye and Pochard is worth it. But here diluted by the masses there is also a good chance of turning up Goosander and Smew.

LEFT: Smew
Smew are very fine birds indeed and we get varying numbers of these winter-visiting saw-billed ducks every year.

Feeding techniques in the mud

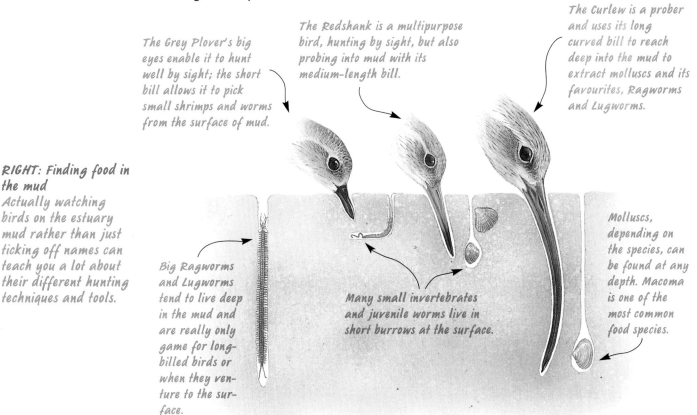

The Grey Plover's big eyes enable it to hunt well by sight; the short bill allows it to pick small shrimps and worms from the surface of mud.

The Redshank is a multipurpose bird, hunting by sight, but also probing into mud with its medium-length bill.

The Curlew is a prober and uses its long curved bill to reach deep into the mud to extract molluscs and its favourites, Ragworms and Lugworms.

RIGHT: Finding food in the mud
Actually watching birds on the estuary mud rather than just ticking off names can teach you a lot about their different hunting techniques and tools.

Big Ragworms and Lugworms tend to live deep in the mud and are really only game for long-billed birds or when they venture to the surface.

Many small invertebrates and juvenile worms live in short burrows at the surface.

Molluscs, depending on the species, can be found at any depth. Macoma is one of the most common food species.

BELOW: Britain's biggest wader
The Curlew is unmistakable, by its size, down-curved bill and bubbling cry.

Minions in the mud

What on Earth are all these birds doing here in the mud of an estuary? To answer this question, get down by the muddy edge of the estuary and poke around a bit. You will find as the birds do which techniques work for different species. In fact, as you play in the mud, try to relate what you are doing to what the birds are doing when you watch them.

For a start you will discover more mud life on a warm day than on a cold one because the animals come to the surface – this also explains why wading birds are much more actively feeding when the air temperature is on the positive side of zero.

Different bird species tend to bias their feeding to certain mud-living organisms, this is reflected in the variety of hardware they have attached to their heads and their spatial distribution within an estuary.

Turn over a forkload of gloopy mud and you will reveal a few centimetres below the surface wonderfully psychedelic red and green Ragworms *Nereis diversicolor*. These are important packages of food for all those waders with mid-length bills. Dig deeper in sandier substrates at the top of the estuary and you will find the fleshy Lugworms *Arenicola marina* so loved by

Curlew. You will also turn up many small molluscs: Cockles and Baltic Tellins. But as you work your way around the estuary it's more or less a question of random searching, a strategy adopted by many wading birds as they stitch their paths across the mudflats, probing until their sensitive bill tip makes contact with food.

This is all very well if you have a long bill, but if you watch short-billed birds such as **Grey** and **Golden Plovers**, you see that the chances of hitting gold with such a limited reach are small. Look at the surface of the mud and you will notice movement – a quiver here and a twitch there. These subtle indiscretions are caused by animals near the surface like *Corophium volutator*, a small crustacean. These animals are so small you would miss them with a fork – in fact the only way to find them is by sight! This is reflected in the feeding tactics of plovers, which have large eyes and feed by making short sprints terminated by a peck.

Duck's dinner

Find out what **Shelduck** are feeding on down at the estuary. Careful where you go, mud can be dangerous – just take a thin skim from the surface of estuary mud. A spoon tied to a long bean pole is ideal and place the mud in a shallow white tray of sea water. Stir it up and allow it to settle. What you should find after a couple of minutes are long spaghetti-like trails in the sediment and at the front of these a tiny little sliding dot that on closer inspection will turn out to be a microscopic snail called the Laver Spire Snail *Hydrobia ulvae*, which feeds on the film of algae that grows on estuarine mud.

ABOVE: Shelducks
Britain's largest duck, the Shelduck is well named – it hoovers up the tiny Laver Spire Shell in the millions with its side-to-side beak sweeping.

LEFT: Lugworm
Dinner? The Lugworm is a bit of a catch for the long-billed birds that are able to reach into their U-shaped burrows in the mud. The squiggle of mud is a sign that these worms are living below.

Invertebrates

When heavy frost turns bare earth to concrete and even the tiniest traces of moisture are chased out into icy crazy paving on our windows and windscreens, winter has arrived. Crunch your way across a frost-kissed landscape, and at first glance it may seem that all has perished. But all around there is life, albeit dangerously close to the edge of the definition.

The more obvious signs, like the rustlings of foraging Blackbirds, the shrill voices of shrews, fresh steaming molehills and the tinkling of mixed flocks of tits and finches as they system-atically work hedgerows, are all testimony to the survival skills of the invertebrate fauna on which they are feeding.

Unlike the famous hibernators, Hedgehogs, dormice and bats, which conserve their reserves by slowing down their metabolism and lowering their body temperature, the invertebrates do not have a body temperature to lower; they are slaves to the temperature of the air.

Insect anti-freeze

Some have the ability to freeze without damag-ing ice crystals forming in their bodies: insect blood has a high concentration of sodium, potassium and chloride ions which basically means it has to drop a lot below zero to freeze. Some produce large amounts of glycerol in their bodies as a form of anti-freeze, which stops ice crystals growing and rupturing their cells and allows the insect to perform super cooling; in some species down to below -50 °C! This is part of the chemical magic that explains the sudden appearance of clouds of dancing **midges**, **Brimstone butterflies** and other winged won-ders during warm winter spells.

Life support capsules

Others create their own life support capsule. There are cocoons and pupae all over the place. Within are held the life of next year's moths, butterflies and flies. **Queen bumble bees** and **wasps** having mated in the late summer find shelter in mouse burrows, under the bark of dead trees or as an alternative in our loft spaces, whilst **Honey Bees** retreat into a tight nucleus of workers surrounding the queen – keeping the nest temperature 20-30 °C higher than ambient by vibrating wing muscles, the whole operation fuelled by a store of honey.

Galls are winter bunkers for the developing **gall wasp grubs** inside. Whilst the icy wind

ABOVE: Blue Tit
The tinkling of mixed flocks of tits and finches as they systematically work hedgerows is a clear sign that invertebrate life is still around in abundance.

BELOW: Small Tortoiseshell and Peacock butterflies
Hibernating butterflies can often be found in your attic or shed.

blasts its way over the downs and through the wild rose bush, tucked up inside the unlikely creation that is a robin's pincushion gall over 15 wasp larvae sit it out.

Leafmould harbour

There are places that do not succumb to the life-numbing cold. Leaves that form drifts at the bottom of hedges are a teeming microcosm – insulated and protected from the elements under a crispy duvet of this year's rejects. Decomposing at ground level are last year's leaves, adding a touch of central heating, keeping the inhabitants of the mulching world above zero and active.

Have a forage yourself and you will see exactly what the Blackbirds are sifting out. As well as the familiar motley assortment of centipedes, beetles, spiders and worms.

Watery sanctuary

Water is another habitat that is buffered against the extremes. Slow to heat up and cool down, it is a surprisingly constant environment and if you need evidence that spring is not the product of spontaneous generation, then look no further than below the surface of your nearest pond or stream. It will have changed in appearance, and even though the floating carpets and streamers of vegetation are not apparent they will be there.

Plants like Water Soldier and Starworts will have simply sunk to the bottom where the water is warmer. Others such as Frogbit, Greater Duckweed and the Milfoils have a neat way of sending off time capsules in the form of nutrient-filled buds, called 'turions', which as the rest of the plant disintegrates, sink to the bottom away from the risk of potential ice damage. Come the spring they float to the surface.

Other aquatic life follows this retreat to the bottom layers. Dip in with a pond net and you will bring to the surface a multitude of invertebrates. You could find **larval** and **nymphal** stages of everything from **damsel-** and **dragonflies**, to various beetles and water bugs.

LEFT: Queen wasp
Queen wasps, having mated in the late summer, find shelter in mouse burrows, under bark or in our homes.

BELOW LEFT: Centipede
They are still everywhere! Soil-living wrigglies like centipedes and millipedes and a plethora of invertebrates can be found when you dig over the garden or allotment – something the Robin knows all too well.

BELOW RIGHT: Dragonfly nymph
The temperature-buffering properties of water make it the perfect nursery. If you don't mind getting cold fingers, there is always wildlife to find, like this dragonfly nymph.

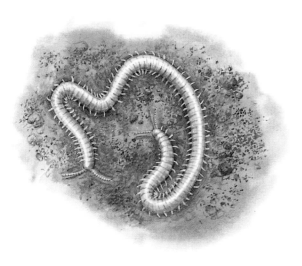

Plants

This is not generally considered to be the best month for a spot of 'botanizing', with most plants reduced to their bare structural bones or hiding in seed, fruit or root stock. But expand your horizons and change your sense of scale and perspective and you will find a plant microcosm of undisputed weirdness. One thing we have plenty of in the winter is moisture; grim grey drizzle rehydrates the lives of the diminutive world of some of this country's strangest organisms – **lichens**.

To appreciate this almost sci-fi world you really need to get down on your hands and knees. Examine just about any surface, tree or stone and you will find them, even in the middle of cities, although due to their extreme sensitivity to air quality, particularly levels of sulphur dioxide, only the toughest survive here.

For the most spectacular a visit to the west or into the highlands and moors is necessary. Many of our upland woods never really lose their pigment but change from greens to russets and, as the leaves fall they take on an almost misty grey-green hue, their branches literally furry with common species such as *Evernia prunastri*. Some branches are festooned with even more bizarre and spectacular growths of species such as **String of Sausages Lichen** *Usnea articulata* and *Usnea florida*. Look a little lower on walls and on tree trunks for the quilted mats of

species such as **Dog Lichen** *Peltigera canina* and **Lungwort** *Lobaria pulmonaria*.

Funky fungi

Although this is not the season traditionally associated with fungi, some of the brightest colours to be seen this month are to be found nestling in the mouldering litter of woods and under hedgerows.

Like split billiard balls with a diameter of a few centimetres, the fruiting bodies of **Scarlet Elf Cups** *Sarcoscypha coccinea* glow from rotten wood! Seen as a blue staining of wood when growing *Chlorociboria aeruginascens* is another even smaller woodland gem. There are many other species with a similar cup-like structure worth looking at under a hand lens as some, like the **Eyelash Fungus** *Scutellinia scutellata* and *Dasyscyphus niveus* (which looks like flattened white cotton pom poms), have exquisite details,

RIGHT: The perfect relationship
Even the unassuming lichens have a neat story. They are a partnership between fungi and algae, which live inside their cells.

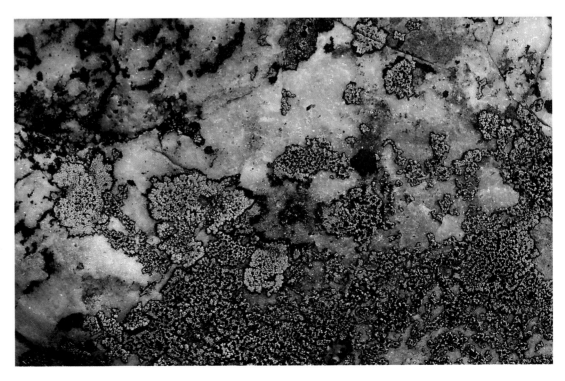

such as fine fringes of hair-like structures. You will have to look a lot closer with a microscope, however, to see the spores, which are held in flimsy envelopes called asci on the surface of the bowl.

As if being of violent coloration isn't enough they match this pigmentation with an explosive finale. The asci, which look like long transparent sausages on their ends, explode, releasing the spores. They always do this towards the light as the tips are positively phototrophic (they follow the light). This ensures the spores are released into the air and not towards the ground. Sometimes they really go out with a bang and they earn themselves the accolade of being probably the only fungus you can hear, as sometimes when the sun hits them they all go off at once – an act known as puffing!

My favourite fungus is the **Jew's Ear Fungus** *Hirneola auriculajudae*. This uncanny brown fungus trembles and feels just like a human ear. It grows on a number of wood types but favours Common Elder. Its curious name is tied in with this host selection. The name in its uncorrupted form would have been Judas Ear, as Judas Iscariot allegedly hung himself from the branches of this very tree. Incidentally, it tastes good as an ingredient in homemade mushroom soup or casserole and can be easily stored dry.

ABOVE: Scarlet Elf-cup Fungus
Like split billiard balls with a diameter of a few centimetres the fruiting bodies of Scarlet Elf-cups (Sarcoscypha coccinea) grow from rotten wood. They can be found in woodlands throughout Britain, but not in large numbers. Elf-cups are not edible!

BELOW LEFT: Dead Man's Fingers
This is one of the few user-friendly common names given to these tiny woodland fungi.

BELOW RIGHT: Jew's Ear Fungus
Go on have a feel! Now you know how this fungus got its name!

Further Reading

Arnold, E N, et al
Field Guide to Reptiles and Amphibians of Britain and Europe
Collins, 1978
ISBN 0 002 19318 3

Baker, Nick
Nick Baker's Bug Book
New Holland Publishers, 2002
ISBN 1 85974 895 3

Beddard, Roy
The Garden Bird Year
New Holland Publishers, 2001
ISBN 1 85974 655 1

Bellman, Heiko
A Field Guide to Grasshoppers and Crickets of Britain and Northern Europe
HarperCollins, 1985
ISBN 0 00219 852 5

Carwardine, Mark
Mark Carwardine's Guide to WhaleWatching
New Holland Publishers, 2003
ISBN 1 84330 059 1

Chinery, Michael
Butterflies of Britain and Europe
Harpercollins and The Wildlife Trusts, 1998
ISBN 0 00220 059 7

Chinery, Michael
A Field Guide to the Insects of Britain and Northern Europe
Harpercollins, 1993
ISBN 0 00219 918 1

Chinery, Micheal
Spiders
Whittet Books, 1993
ISBN 1 87358 009 6

Corbet, G., Southern H., Eds
The Handbook of British Mammals
Blackwell, 1964
ISBN 0 632 09080 4

Fitter, R., Manuel, R.
Field Guide to the Fresh Water Life of Britain and North-west Europe
Collins, 1986
ISBN 0 002 19143 1

Gibbons, B.
Dragonflies and Damselflies of Britain and Northern Europe
Country Life Books, 1986
ISBN 0 600 35841 0

Gibbons, B.
Field Guide to the Insects of Britain and Northern Europe
The Crowood Press, 1996
ISBN 1 85223 895 X

Golley, Mark
Birdwatcher's Pocket Field Guide
New Holland Publishers, 2003
ISBN 1 84330 119 9

Golley, Mark
The Complete Garden Bird Book
New Holland Publishers, 2001
ISBN 1 84330 035 4

Harde, K W
Beetles
Bookmart, 1999
ISBN 1 856 05448 9

Hammond, Nicholas (Series Editor)
The Wildlife Trusts Guide to Birds
New Holland Publishers, 2002
ISBN 1 85974 958 5

Hammond, Nicholas (Series Editor)
The Wildlife Trusts Guide to Butterflies and Moths
New Holland Publishers, 2002
ISBN 1 85974 959 3

Hammond, Nicholas (Series Editor)
The Wildlife Trusts Guide to Garden Wildlife
New Holland Publishers, 2002
ISBN 1 85974 961 5

Hammond, Nicholas (Series Editor)
The Wildlife Trusts Guide to Insects
New Holland Publishers, 2002
ISBN 1 85974 962 3

Hammond, Nicholas (Series Editor)
The Wildlife Trusts Guide to Trees
New Holland Publishers, 2002
ISBN 1 85974 965 8

Hammond, Nicholas (Series Editor)
The Wildlife Trusts Guide to Wild Flowers
New Holland Publishers, 2002
ISBN 1 85974 966 6

Hammond, Nicholas (Series Editor)
The Wildlife Trusts Handbook of Garden Wildlife
New Holland Publishers, 2002
ISBN 1 85974 960 7

Jonsson, Lars
Birds of Europe with North Africa and the Middle East
Helm, 1999
ISBN 0 713 65238 1

Leach, M.
Mice of the British Isles
Shire Publications, 2000
ISBN 0 747 80056 1

Moss, Stephen and Cottridge, David
Attracting Birds to your Garden
New Holland Publishers, 2000
ISBN 1 85974 005 7

Moss, Stephen
Birdwatcher's Guide: How to Birdwatch
New Holland Publishers, 2003
ISBN 1 84330 154 7

Nancarrow, L., Hogan Taylor, J.
The Worm Book
Ten Speed Press, 1998
ISBN 0 89815 994 6

North, Ray
Ants
Whittet Books, 1996
ISBN 1 87358 025 8

Oddie, Bill
Bill Oddie's Birds of Britain and Ireland
New Holland Publishers, 1998
ISBN 1 85368 898 3

Oxford, R.
Minibeast Magic – Kind-hearted Capture Techniques for Invertebrates
A Yorkshire Wildlife Trust Publication, 1999
ISBN 9 780950 946020

Packham, Chris
Chris Packham's Back Garden Nature Reserve
New Holland Publishers, 2001
ISBN 1 85974 520 2

Porter, Jim
The Colour Identification Guide to Caterpillars of the British Isles
Viking, 1997
ISBN 0 67087 509 0

Powell, Dan
A Guide to the Dragonflies of Great Britain
Arlequin Press, 1999
ISBN 1 90015 905 8

Rackham, Oliver
The History of the Countryside
J.M. Dent & Sons Ltd, 1986
ISBN 0 460 86091 7

Robert, Michael J.
The Collins Field Guide to the Spiders of Britain and Northern Europe
HarperCollins, 1995
ISBN 0 00219 981 5

Skinner, B.
Colour Identification Guide to Moths of the British Isles
Viking, 1984
ISBN 0 67080 354 5

Skinner, G.
Ants of the British Isles
Shire Publications, 2000
ISBN 0 85263 896 5

Thomas, Jeremy; Lewington, Richard
The Butterflies of Britain and Ireland
Dorling Kindersley, 1991
ISBN 0 86318 591 6

Wardhaugh, A. A.
Bats of the British Isles
Shire Publications, 2000
ISBN 0 74780 303 X

Wardhaugh, A. A.
Land Snails of the British Isles
Shire Publications, 2000
ISBN 0 74780 027 8

Wisniewski, Partick J.
Newts of the British Isles
Shire Publications, 2000
ISBN 0 747 80029 4

Series of Naturalists' Handbooks
Richmond Publishing Co. Ltd.

Useful Addresses

The Wildlife Trusts
The Kiln
Waterside
Mather Road
Newark NG24 1WT
Tel: 0870 0367711
Fax: 0870 0360101
Email: info@wildlife-
trusts.cix.co.uk
Web: www.wildlifetrusts.org

Wildlife Watch
as for The Wildlife Trusts
Email: watch@wildlife-
trusts.cix.co.uk
Web: www.wildlife-watch.org

**The Amateur Entomologists'
Society**
PO Box 8774
London SW7 5ZG
Email: aes@theaes.org
Web: www.theaes.org

Bat Conservation Trust
15 Cloisters House
8 Battersea Park
London SW8 4BG
Tel: 020 7627 2629
Email: Enquiries@bats.org.uk
Web: www.bats.org.uk

**Bees, Wasps and Ants
Recording Society**
Nightingales
Haslemere Road
Milford
Surrey GU8 5BN
Web: website.lineone.net/
~ammophila/

British Arachnological Society
Secretary: Dr Helen J. Read
2 Egypt Wood Cottages
Egypt Lane
Farnham Common
Bucks SL2 3LE
Tel: 01753 645791
Email: secretary@
britishspiders.org.uk
Web: http://www.salticus.
demon.co.uk

**British Butterfly Conservation
Society**
Manor Yard
East Lulworth
near Wareham
Dorset BH20 5QP
Tel: 01929 400209
Email: info@butterfly-
conservation.org
Web: www.butterfly-
conservation.org

British Dragonfly Society
The Haywain
Hollywater Road
Bordon
Hampshire
GU35 OAD
Email: bdswebmaster@
hanslope.demon.co.uk
Web: www.dragonflysoc.
org.uk

**British Trust for Conservation
Volunteers (BTCV)**
36 St Mary's Street
Wallingford
Oxfordshire
OX10 0EU
Tel: 01491 839766
Fax: 01491 839646
Email: Information@
btcv.org.uk
Web: www.btcv.org

**British Trust for Ornithology
(BTO)**
The Nunnery
Thetford
Norfolk
IP24 2PU
Tel: 01842 750050
Fax: 01842 750030
Email: general@bto.org
Web: www.bto.org

**Brunel Microscopes (BR)
Limited**
Unit 12 Enterprise Estate
Chippenham
Wilts
SN14 6QA
Tel. 01249 462 655
Fax 01249 445 156
Email: brunelmicro@
compuserve.com
Web: www.brunelmicroscopes.
co.uk

**Buglife – The Invertebrate
Conservation Trust**
200 Salisbury Road
Totton
Southampton
SO40 3PE

**Conchological Society of Great
Britain and Ireland**
35 Bartlemy Road
Newbury
Berks RG14 6LD
Tel: 01635 42190
Fax: 01635 820904
Email: membership
@conchsoc.org
Web: www.conchsoc.org

Field Studies Council
Preston Montford
Montford Bridge
Shrewsbury SY4 1HW
Tel: 01743 850674
Fax: 01743 852101
Email: fsc.headoffice@
ukonline.co.uk
Web: www.field-studies-
council.org

Froglife
Mansion House
27-28 Market Place
Halesworth
Suffolk
IP19 9AY
Tel: 01986 873733
Fax: 01986 874744
Email: froglife@froglife.
demon.co.uk
Web: www.froglife.org

**Herpetofauna Conservation
Trust**
655a Christchurch Toad
Boascombe
Bournemouth
Dorset BH1 4AP
Tel: 01202 391319
Fax: 01202 392785
Email: HerpConsTrust@hcontrst.
force9.net
Web: www.hcontrst.f9.co.uk

The Mammal Society
15 Cloisters House
8 Battersea Park Road
London SW8 4BG
Tel: 020 7498 4358
Fax: 020 7622 8722
Email: enquiries@mammal.org.uk
Web: www.abdn.ac.uk/mammal/

The Microscope Shop
Oxford Road
Sutton Scotney
Winchester
SO21 3JG

Oxford Bee Company Ltd
40 Arthur Street
Loughborough
Leics
LE11 3AY
Tel: 01509 261654
Plantlife
21 Elizabeth Street
London
SW1W 9RP
Tel: 020 7808 0100
Fax: 020 7730 8377
Email: enquiries@plantlife.org.uk
Web: www.plantlife.org

**Royal Society for the Protection
of Birds (RSPB)**
The Lodge
Sandy
Bedfordshire
SG19 2DL
Tel: 01767 680551
Fax: 01767 692365
Email: bird@rspb.demon.co.uk
Web: www.rspb.org.uk

Small Life Supplies
Station Buildings
Station Road
Bottesford
Notts
NG13 OEB
Tel: 01949 842446
Fax: 01949 843036
Web: www.small-life.co.uk

**Watkins and Doncaster the
Naturalists**
PO Box 5
Cranbrook
Kent
TN18 5EZ
(General naturalist supplies –
nets, boxes, moth traps, pooters,
etc.)
Tel: 01580 753133
Fax: 01580 754054
Email: robin.ford@virgin.net
Web: www.watdon.com

**Whale and Dolphin
Conservation Society**
P.O. Box 232
Melksham
Wiltshire
SN12 7SB
UK
Tel: 0870 870 5001
Fax: 0870 870 5002
Email: info@wdcs.org
Web: www.wdcs.org

Wildfowl and Wetlands Trust
Slimbridge
Gloucestershire
GL2 7BT
Tel: 01453 891900
Fax: 01453 890827
Email: enquiries@wwt.org.uk
Web: www.wwt.org.uk

Index

Page numbers in **bold** refer to illustrations and in *italics* refer to maps

Acknowledgements

Lots of naturalists that I have met over the years of my life so far have contributed in one way or another to this book – most without even knowing it! The best bit about the natural world is that it is everywhere and you will never be bored. It is also so intricate and fascinating down to the minute details. There will always be questions. I would like to thank all the people who have answered my questions – whether they were people I met in a bird hide, out for a walk, or while I've been making wildlife programmes for radio and TV.

In particular, it cannot be easy raising a naturalist, so a big thank you to my parents Sandy and Steve, my brother Paul and my sadly missed grandparents, who took me on rambles and showed me sticklebacks, frogs, birds and Winkles!

Also, a big 'thanks very much' to all the staff at BBC Wildlife Magazine, especially Roz, Jane and Alex, who gave me a big break in the first place.

I would also like to thank all those at New Holland, especially Jo Hemmings, for encouraging me to write this book, and to Lorna Sharrock for not shouting at me and getting too cross when I gave her some pathetic excuse for not meeting a deadline!

My neighbours Rob, Julia, Madeleine and Joe for allowing me and photographer Dave Cottridge to invade their pond!

Nick Baker

Photographic acknowledgements
All photographs by David M. Cottridge, with the exception of the following:
Ardea: Bob Gibbons: p122(t); Stefan Meyers: p38(b); P. Morris: p123(b)
Bruce Coleman Collection: Kim Taylor: p140
Chris Gomersall: p133
Nature Photographers Ltd: S.C. Bisserot: pp44, 45(tr), 81, 98(bl), 99(bl); Brinsley Burbidge: pp4(tr), 94(b); Robin Bush: pp58(br), 89(t); N.A. Callow: p91; Colin Carver: pp54(tr), 64(t), 137(t), 142(t); Hugh Clark: pp4(tc), 73(b); Andrew Cleave: pp45(tl), 151(b); Phil Green: p128(bl); Jean Hall: p112; Michael J. Hammett: pp65(t), 102(t); E.A. Janes: pp46(b), 60(t), 73(t); Charles Palmar: p139; Don Smith: p148; Paul Sterry: back cover (c, r), pp4(br), 5(tc), 9, 13(l), 34(t), 50(b), 58(t), 64(b), 65(b), 71(tl), 79(b), 80(t), 84(t), 94(t), 96(b), 100(b), 103(tr), 105, 106(t), 107(b), 111, 123(t), 142(b), 144(t), 145(t), 155(br); Roger Tidman: p74
Richard Revels: cover spine (t), pp1, 4(bl), 5(tl, bl), 12, 22, 41, 46(t), 47, 49, 51, 58(bl), 68(t), 89(b), 93, 99(br), 101, 102(b), 103(tl), 104, 106(b), 113, 119(t,b), 122(b), 128(br), 141, 143(t, b), 146(b), 155(bl)
Alan Williams: pp26, 40(tr), 53, 63, 77(br), 120
Windrush Photos: David Tipling: pp4(b), 126(b), 132(t), 137(b); Marcus Young: p40(tl)

Artwork acknowledgements
All artwork by Wildlife Art Ltd, with the exception of the following:
David Daly: pp15, 17, 18(b), 21, 31, 32, 33, 37(br), 39, 40, 53, 61, 87, 88, 108, 121, 133, 134, 135, 136, 138, 147, 148
Sheila Hadley: pp24, 36, 37(t), 49, 101, 116, 128, 131
Stephen Message: pp18(t), 62, 74, 75, 77, 98, 109
Mike Unwin: p132

(t= top; b=bottom; c=centre; l=left; r=right)